University of California Publications

GEOLOGICAL SCIENCES
Volume 121

Contributions to the Neogene Paleobotany of Central California

by Daniel I. Axelrod

University of California Press

CONTRIBUTIONS TO THE NEOGENE PALEOBOTANY

OF CENTRAL CALIFORNIA

Contributions to the Neogene Paleobotany of Central California

by Daniel I. Axelrod

UNIVERSITY OF CALIFORNIA PRESS
Berkeley • Los Angeles • London

UNIVERSITY OF CALIFORNIA PUBLICATIONS IN GEOLOGICAL SCIENCES

Editorial Board: D. I. Axelrod, W. B. N. Berry, R. L. Hay, M. A. Murphy,
J. W. Schopf, W. S. Wise, M. O. Woodburne

Volume 121

Issue Date: December 1980

UNIVERSITY OF CALIFORNIA PRESS
BERKELEY AND LOS ANGELES
CALIFORNIA

UNIVERSITY OF CALIFORNIA PRESS, LTD.
LONDON, ENGLAND

Library of Congress Cataloging in Publication Data

Axelrod, Daniel I.
   Contributions to the Neogene paleobotany of
central California.

   (University of California publications in
geological sciences; v. 121)
   1. Paleobotany—Tertiary. 2. Paleobotany—
California. I. Title. II. Series: California.
University. University of California publications
in geological sciences; v. 121.
QE926.A95   560′.1′78   80-15355
ISBN 0-520-09621-5

Contents

*A detailed table of contents appears
with each numbered chapter.*

Acknowledgments

This research has been sponsored by a grant from the National Science Foundation (Grant GB-37533), which is acknowledged with thanks. Prof. Cordell Durrell has kindly identified the diverse andesites of the Mt. Reba area, and Harold Bonham selected samples suitable for radiometric dating. Thanks are also extended to Prof. Harry P. Bailey for permission to use the thermal nomogram that he constructed. When used to chart modern temperatures, it provides a more reliable basis for estimating the paleo-temperatures under which the fossil floras lived. Finally, it is a pleasure to thank Sandra Chase for drafting the figures.

Abstract

This volume describes three new fossil floras from central California that contribute to an understanding of diverse ecologic, climatic, and tectonic problems.

The Mt. Reba flora (7 m.y.) now occurs at 8,700 ft. (2,650 m), close to timberline in the Sierran summit region. Its taxa contributed to vegetation similar to the Douglas fir and mixed evergreen forests now in the middle-upper foothill belt of the Sierra near 2,500-3,000 ft. (760-915 m), at the lower margins of the dominant mixed conifer forest. The estimated difference in mean annual temperature of about 22°F (12.2°C) between the fossil flora and that now at the locality implies that the area has been elevated about 6,000 ft. (1,830 m) since 5-7 m.y. ago. The data lead to a revision of previous estimates of rates of erosion in the Sierra, and of the ages of the Mountain Valley and Canyon stages of erosion.

The Turlock Lake flora (4.5 m.y.) is now at an elevation of 240 ft. (73 m) in open grassland on the lowest Sierran slope. Its species are similar to those now in the upper oak woodland belt, close to mixed evergreen forest. The flora indicates a wetter climate than the nearby, slightly older ($\sim$ 6 m.y.) Oakdale flora. The rise in precipitation corresponds to the start of buildup of major icecaps in the boreal region, and to the initiation of glaciation in the high Sierra shortly thereafter (3.0-2.7 m.y.).

The Broken Hill flora ($\sim$ 5 m.y.) from the basal San Joaquin Formation near Kettleman City shows that this presently semi-desert area then supported oak-woodland vegetation like that now on the inner slopes of the Santa Lucia Mountains. The assemblage reflects a mild, warm climate favorable for abundant *Persea* and other Tertiary relicts, notably *Magnolia, Sapindus, Ulmus*, and two species of *Populus*.

Comparison of these Hemphillian-age floras with others in central California and western Nevada reveals marked differences in composition over moderate distances, reflecting local terrain and climate. The fossil communities were not as varied as those of today, which results from the continuing trend to greater differences in local terrain and climate. The mid-Hemphillian period of drought, which is recorded in central California, western Nevada, and the High Plains, rapidly modernized the surviving flora that gave rise to our modern ecosystems. This period of greatest aridity during the Tertiary may reflect the spread of the Antarctic icecap about 5-6 m.y. ago.

I

INTRODUCTION

Chapter I Contents

INTRODUCTION

During the past decade, the rise of plate tectonics as a new subscience of geology has greatly illuminated our understanding of earth history, notably the physical processes that account for the spreading of the ocean floor and rafting of crustal plates to new positions. This has led to a better understanding of tectonic history on the continents, and has also clarified major problems of evolution and biogeography, both on land and in the sea.

Attention is directed here to evidence regarding the nature of environmental change during the Later Neogene, a time when two major conditions resulting largely from tectonic events are recorded. One was the pronounced period of drought during the Middle Hemphillian (= Messinian), a time when increased aridity is recorded in widely separated areas (Axelrod, 1944, p. 215; 1948, 1971; Adams et al., 1977; Hsu et al., 1977). The second was Neogene episodes of glaciation, as inferred from paleobotanical evidence (Axelrod, 1944, 1971), the underlying causes of which lie largely in tectonic history (Adams et al., 1977; Ingle, 1967; Kennett, 1977; Hallam, 1976). A third event of more local nature, but of wide implication, was the time of major uplift of the Sierra Nevada, and hence the rate at which a root of sufficient size was created to support the newly elevated region (Axelrod, 1957). New data that bear on these problems are provided by three previously undescribed fossil floras in central California (Fig. 1): the Mt. Reba flora from the central Sierran summit region; the Turlock Lake flora from the low Sierran piedmont; and the Broken Hill flora from the edge of the Central Valley bordering the inner south Coast Ranges.

AGE ASSIGNMENT

During the extensive and continuing worldwide gathering of new data from the sea floor by *Glomar Challenger*, numerous deep-sea cores have been recovered and dated by their radiometric and paleomagnetic properties and by their fossil content. As a result, many well-dated local sequences have been established, the history of the ocean floor has been clarified, and paleontologic correlations have been refined over wide regions (e.g., La Brecque, Kent, and Cande, 1977; Berggren and Van Couvering, 1974; Haq, Berggren, and Van Couvering, 1977).

One of the important outcomes of these continuing, voluminous studies (e.g., Kasbeer, 1973, 1975) has been revision of the time-span of the classical ages and stages of the Cretaceous and Tertiary periods (Cohee et al., 1978). With respect to the present study, the revised dates for the type Miocene and Pliocene epochs involve shifting the Miocene boundary upward from 12 (or 10) to 5 million years (m.y.), with the resultant shortening of the Pliocene Epoch to a span of only 3.5 m.y. This is not a new notion, but was discussed a quarter-century ago (e.g., Stirton, 1951, p. 77). In any event, this usage has been quickly adopted by marine invertebrate (microfossil) paleontologists, and is used also as the standard reference for dating plate tectonic events. The upward restriction of

5

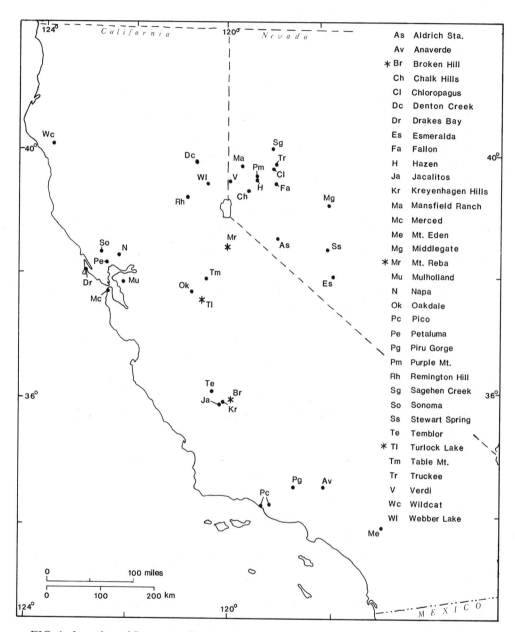

FIG. 1. Location of floras described in this report are marked by a star (*), and others of Late Neogene age are also indicated.

the Pliocene to the time from 5 to 1.5 m.y. means that all floras that were described earlier as Early and Middle Pliocene now become Late Miocene in terms of current, revised Lyellian Age usage. However, this does not modify their actual age as determined either by radiometric or mammalian (evolutionary) evidence (Table 1).

In the larger view, it is perhaps less important to refer a flora to the Miocene or Pliocene than to indicate, when possible, its age in actual time (millions of years). Readers who may relate this study to earlier ones that describe the Late Tertiary floras of California and other areas will therefore have to adjust to the new Lyellian age assignments if they choose to use them. For the sake of clarity and continuity with past work, I shall refer to the age of floras in terms of million years (m.y.) or to established mammalian stages which also date the rocks (Evernden et al., 1964). Since numerous mammalian taxa ranged across the region through varied vegetation-climatic belts, and since they were evolving at relatively rapid rates, they provide a more reliable basis for age reference than the plants. Inasmuch as the floras often are associated with fossil mammals, assignment of the floras to provincial mammalian ages seems justified. In this regard, the provincial floral ages based on plant sequences in Alaska (Wolfe, Hopkins, and Leopold, 1966; Wolfe, 1966) cannot be applied in this area, because the taxa are not here: a stage or zone can be recognized away from its type area only on the basis of the occurrence of its taxa in other areas. The limited value of plant age-stage terminology becomes especially apparent in the regional discussion presented below, in which the marked changes in the floras of Hemphillian Age (9-4 m.y.) from the central coast of California across the low Sierran rise into western Nevada are described.

OUTLINE OF FLORAS

The floras described in the following pages can be characterized briefly in terms of their location, geologic occurrence, composition, and age as follows.

Mt. Reba flora

Located in the upper subalpine belt at an elevation of 8,700 ft. (2635 m), on the crestal ridge 1 mile east of Mt. Reba, overlooking the 4,500-foot-deep Mokelumne River gorge.

Preserved in andesitic sandstones and mudflow breccias that represent the Disaster Peak Formation.

A total of 14 species, distributed among 7 conifers, 1 monocot, and 6 dicots, represented by 1,024 specimens.

Middle Hemphillian in age, dated at 7 m.y., the average of K/Ar age determination by three different laboratories.

Turlock Lake flora

On islands in Turlock Lake Reservoir, situated in the lowest Sierran foothills east of Modesto at an elevation of 240 ft. (73 m).

Preserved in buff-colored mudstones intercalated with andesitic sandstone and pebble conglomerate of the upper Mehrten Formation.

A total of 25 species, distributed among 1 conifer, 4 monocots, and 20 dicots, represented by 415 specimens.

Associated with a large Late Hemphillian mammalian fauna, about 4.5 m.y. in age.

M.Y.	EPOCH	STAGE	SIERRA NEVADA	WESTERN GREAT BASIN	CENTRAL CALIFORNIA	SOUTHERN CALIFORNIA
	Pleisto-cene	Rancholabrean				
		Irvingtonian				
					Santa Clara	Soboba
2			Owens Gorge	Coso		
	Pliocene	Blancan	San Joaquin Mt.			
				Wichman	Sonoma Napa	
					Kreyenhagen Hills	
4			Turlock Lake			
					Broken Hill	
			Oakdale	Verdi		Mt. Eden
6			Mt. Reba	Hazen	Petaluma	Piru Gorge
	Miocene	Hemphillian			Mulholland	
				Alturas		Anaverde
8				Chalk Hills		
10			Remington Hill		Black Hawk Ranch	
		Clarendonian			Diablo	Ricardo
					Neroly	
12			Table Mt.	Aldrich Sta.		Mint Canyon
				Fallon		

TABLE 1. Ages of Late Neogene Floras in the California region and the Western Great Basin.

Broken Hill flora

At the south end of North Dome, Kettleman Hills, situated west of Kettleman City at an elevation of 715 ft. (218 m).

Preserved in blue-gray andesitic sandstones and shales of the Cascajo Member, the basal unit of the San Joaquin Formation.

A total of 22 species, all dicots, represented by 238 specimens.

Late Hemphillian age is indicated by its position well below Blancan-age mammals and by a radiometric date of 4.5 m.y. on a tuff above the flora, which is thus inferred to be about 5 m.y.

REFERENCES CITED

Adams, C.G., R.H. Benson, R.B. Kidd, et al.
   1977   The Messinian salinity crisis and evidence of late Miocene eustatic changes in the world ocean. Nature 269: 383-386.
Axelrod, D.I.
   1944   The Pliocene sequence in central California. In R.W. Chaney (ed.), Pliocene Floras of California and Oregon. Carnegie Inst. Wash. Pub. 553: 207-224.
   1948   Climate and evolution in western North America during Middle Pliocene time. Evolution 2: 127-144.
   1957   Paleoclimate as a measure of isostasy. Amer. Jour. Sci. 255: 690-696.
   1971   Fossil plants from the San Francisco Bay region. In Geologic Guide to the Northern Coast Ranges, Point Reyes Region, California. Geol. Soc. Sacramento, Field Trip Guidebook for 1971, pp. 74-86.
Berggren, W.A., and J.A. Van Couvering
   1974   Biostratigraphy, geochronology, and paleoclimatology of the last 15 million years in marine and continental sequences. Paleogeography, Paleoclimatology, Paleoecology 16: 1-216.
Cohee, G.V., M.F. Glaessner, and H.H. Hedberg
   1978   The Geologic Time Scale. Amer. Assoc. Petrol. Geol., Studies in Geology No. 6. 388 p.
Evernden, J.F., D.E. Savage, G.H. Curtis, and G.T. James
   1964   Potassium-argon dates and the Cenozoic mammal chronology of North America. Amer. Jour. Sci. 262: 145-198.
Hallam, A.
   1976   Antarctic ice and desiccation in the Mediterranean. Nature 263: 194.
Haq, B.W., W.A. Berggren, and J.A. Van Couvering
   1977   Corrected age of the Pliocene-Pleistocene boundary. Nature 269: 483-488.
Hsu, K.J., et al.
   1977   History of the Mediterranean salinity crisis. Nature 267: 399-403.
Ingle, J.C., Jr.
   1967   Foraminiferal biofacies variation and the Miocene-Pliocene boundary in southern California. Bulletin of Amer. Paleontology 52: 217-394.
Kasbeer, T.
   1973   Bibliography of continental drift and plate tectonics, Vol. I. Geol. Soc. Amer. Spec. Paper 942. 96 pp.
   1975   Bibliography of continental drift and plate tectonics, Vol. II. Geol. Soc. Amer. Spec. Paper 175. 151 pp.

Kennett, J. P.
  1977 Cenozoic evolution of Antarctic glaciation, the Circum-Antarctic Ocean and their impact on global paleooceanography. Jour. Geophys. Res. 82: 3843-3959.
La Brecque, J.L., D.V. Kent, and S.C. Cande
  1977 Revised magnetic polarity time scale for Late Cretaceous and Cenozoic time. Geology 5: 330-335.
Stirton, R.A.
  1951 Principles in correlation and their application to later Cenozoic Holarctic continental mammalian faunas. Internat. Geol. Congr., 18th Sess., Great Britain, pt. XI: 74-84.
Wolfe, J.
  1966 Tertiary plants from the Cook Inlet region, Alaska. U.S. Geol. Surv. Prof. Paper 398-B: B1-B32.
Wolfe, J.A., D.M. Hopkins, and E.B. Leopold
  1966 Tertiary stratigraphy and paleobotany of the Cook Inlet region, Alaska. U.S. Geol. Surv. Prof. Paper 398-A: A1-A29.

II

THE MT. REBA FLORA
FROM ALPINE COUNTY

Chapter II Contents

TABLES

PLATES

INTRODUCTION

Up to the present time, the described Late Neogene floras from the Sierra Nevada have included only the Oakdale flora (Axelrod, 1944b) from the western foothill belt and the Verdi (Axelrod, 1958) at the eastern base of the range (see Fig. 1). The recent discovery of the Mt. Reba flora at a site high up in the subalpine zone in Alpine County thus provides our initial information concerning the nature of vegetation and climate in the higher parts of the range during Hemphillian time. This has a critical bearing on understanding the history of the ecosystems that are now there. The Mt. Reba flora can also add to our understanding of the topographic history of the Sierra Nevada in this area, because paleoecologic data make it possible to estimate the altitude of the crestal area.

The presence of fossil plants near Mt. Reba was brought to my attention by Mrs. Helen Eaton of Davis, California, to whom thanks are extended. It is also a pleasure to acknowledge the assistance of Donald E. Stikkers and Donald C. Kessler of the U.S. Forest Service, Arnold, California, in securing a permit to work in the area. Maury Rasmussen of Bear Valley generously made his caterpillar tractor available for excavating, and John Burgess operated it, skillfully exposing the plant-bearing andesitic sandstones.

PRESENT PHYSICAL SETTING

The Mt. Reba flora occurs high up on the west slope of the central Sierra Nevada, 12 miles (19.2 km) west of Ebbetts Pass and 2 miles (3.2 km) north of Lake Alpine. The site is at an elevation of 8,650 ft. (2,635 m) one mile east of Mt. Reba (alt. 8,758 ft.; 2,669 m), on the ridge that forms the south wall of Underwood Valley (Plate 1, figs. 1 and 2). The immediate area is drained by tributaries of the Mokelumne River, which is in a gorge 2.5 miles (4 km) northwest, and 4,000 ft. (1,200 m) below, the fossil site. A 360° view from the summit ridge where the flora occurs reveals that the west flank of the Sierra slopes gently to the Great Valley 70 miles (113 km) distant. It also shows the nearly even surface of the crystalline basement on which are perched the varied Tertiary volcanic and sedimentary rocks that blanket the interfluves today. Small, local volcanic centers in the Sierran summit area 10-12 miles east, notably Reynolds Peak and Raymond Peak, reach elevations of 9,700 ft. (2,957 m) and 10,000 ft. (3,048 m) respectively.

The locality is above timberline, which in this local area is determined primarily by high winds and the shallow, pervious, dry substrate in which the plants live. In more favorable situations in the nearby area, timberline normally is close to 9,300 ft. (2,835 m). The area near timberline is dominated chiefly by low shrubs, including *Artemisia tridentata*, *Amelanchier utahensis*, *Holodiscus dumosus*, *Ribes cereum*, and *Symphoricarpos oreophilus*, and various perennials, notably *Allium*, *Castellija*, *Lewisia*, *Lupinus*, *Phacelia*, *Penstemon*, and *Wyethia*. The forest is made up of dense stands of *Abies magnifica*, *Pinus monticola*, and *Tsuga mertensiana* which attain heights of over 100 ft.

15

(30 m) in sheltered sites and regularly form a mixed forest. *Abies concolor, A. magnifica,* and *Pinus jeffreyi* are the chief forest dominants at altitudes below 8,000 ft. (2,438 m). *Pinus murrayana* commonly forms dense stands on moister, cooler flats in valleys, and thickets of *Populus tremuloides* occur throughout the region. Associated shrubs in the subalpine forest zone include *Acer glabrum, Amelanchier alnifolia, Arctostaphylos nevadensis, Castanopsis sempervirens, Ceanothus cordulatus, Quercus vaccinifolia, Ribes viscosissimum,* and *Sorbus scopulina.*

Climate at the fossil site is generally similar to that at Twin Lakes (elev. 7,823 ft; 2,385 m) in the area 14 miles (23 km) north, and at Tamarack (elev. 8,060 ft.; 2,457 m), situated 8.5 miles (13.7 km) northeast, close to the Sierran summit near Lower Blue Lake. Both meteorological stations are in the red fir-white pine-mountain hemlock forest zone. Twin Lakes has a mean annual temperature of 38.9° F (3.8°C), Tamarack of 38.6° F (3.7°C). The annual range of mean monthly temperature is 32° F (17.8°C) at Twin Lakes and 31.1°F (17.3°C) at Tamarack. These figures indicate that the warmth (*W = ET,* or *effective temperature*) of climate is *W* 51.5° F (10.8°C) at Twin Lakes, and *W* 51.3°F (10.7°C) at Tamarack, or that only 70 to 76 days of the year have a mean temperature above those levels (see Bailey, 1960).

Equability (*M* = temperateness) rating is *M* 42.3 at these stations; by contrast, San Francisco, which is situated 140 miles (225 km) southwest, has a rating of *M* 74 on a scale of 100 (see Bailey, 1964). Temperature data for Twin Lakes indicate that timberline (marked by *W* 50°F, or 10°C) is 1,600 ft. (485 m) higher. This agrees with the distribution of vegetation in that area, for on Elephants Back, situated 4 miles (6.4 km) south-southeast of Twin Lakes, timberline is at 9,300 ft. (2,835 m). Precipitation averages 46.4 in. (1,179 mm) at Twin Lakes, and 51.7 in. (1,313 mm) at Tamarack, with most of it falling as snow during the period from late October into late March. Summers are very dry, with only occasional showers along the crestal area in July and August. Figure 2 charts the average mean monthly temperature and precipitation at these stations. Similar conditions prevail at the Mt. Reba fossil site, though its position about 800 ft. (244 m) higher gives it somewhat lower temperatures, as estimated below in the section on climate.

GEOLOGY

Stratigraphy

The stratigraphic occurrence of the Mt. Reba flora was established by mapping the geology of the ridge eastward from Mt. Reba for about 2 miles, and the adjacent areas of Underwood Valley to the north and Lake Alpine to the south. As shown on Plate 1 and also in Figures 3 and 4, the flora occurs near the top of the Tertiary sequence which rests unconformably on the Sierra crystalline basement.

Basement

The oldest part of the crystalline basement in this area is a metamorphic pendant exposed near Mt. Reba Ski Bowl and on the south slope of Mt. Reba. It widens as it trends northwest to include the major eminence of Mokelumne Peak (9,300 ft.; 2,835 m),

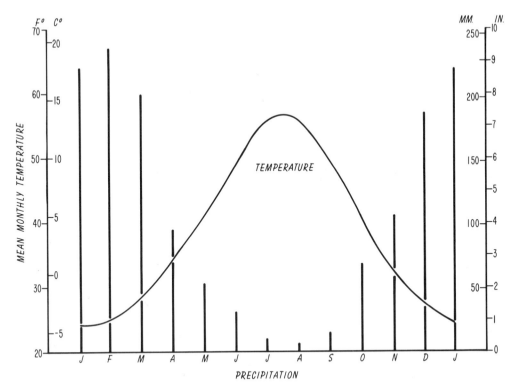

FIG. 2. Mean monthly temperature and precipitation data for two stations near the Mt. Reba locality. Elevation at Tamarack (8-yr. record) is 8,060 ft. (2,457 m), at Twin Lakes (19-yr. record) 7,823 ft. (2,385 m). Both stations are in the subalpine forest of *Abies magnifica, Tsuga mertensiana,* and *Pinus monticola* that also dominates the Mt. Reba area, which has lower winter and summer temperatures because it is at a slightly higher (8,758 ft.; 2,669 m) elevation.

across the Mokelumne River gorge. As interpreted by Gilbert (1959), the pendant is composed of nearly vertical-dipping metavolcanic and metasedimentary rocks which are folded about a steep axis that strikes northwest. The metamorphic rocks include meta-andesite altered to hornblende-biotite hornfels, metapyroclastic biotite schist locally altered to hornblende diopside hornfels, and mica-rich metasedimentary rocks altered to biotite schist and biotite hornfels. The least metamorphosed rocks, which have relict textures, are in the north part of the pendant near a pluton of granodiorite. The most highly crystallized rocks, altered by injection and permeation of granitic material, are in the central and southern parts of the pendant and are related to a younger pluton of quartz monzonite porphyry (Gilbert, 1959).

The pendant may represent highly metamorphosed parts of the Paleozoic Calaveras Formation that crops out at lower levels to the west. Since it covers only a small part of the area studied, it was included with the granitic rocks in mapping. As viewed from the summits of the higher ridges (Plates 1 and 2), the basement of granodiorite and allied rocks extends for many miles in every direction, and was essentially a peneplain at the time of the earliest volcanic eruptions.

AGE		FORMATION	SYMBOL	THICK-NESS (m)	LITHOLOGY	DESCRIPTION
QUATERNARY		*Alluvium*	Qal – Qt	0 – 30		Valley alluvium (Qal), glacial till (Qt).
PLIOCENE		*Mt. Reba cgl.*	Tcg	0 – 20		Giant boulder cgl., chiefly andesite, with very large clasts.
7 m.y.→		*Disaster Pk.*	Td	0 – 10		Thin hbl. andesite mudflow breccias, interbedded andesitic ss. with rolled fossil plants. (K/Ar = 7 ± 1 m.y.)
11 m.y.→ M I O C E N E		*Relief Peak*	Tr	0 – 100		Pink, brown and gray hbl. hypersthene andesite, hypersthene andesite mudflow breccias and intrusions. Thin local andesite boulder cgl. and ss. Intruded by black hbl. andesite. (K/Ar = 11 m.y.)
		Bear Valley	Tb	0 – 245		Gray to brown cgl. and ss., mostly andesite clasts, with some basement debris chiefly in basal part. Locally interbedded with Underwood mudflow breccias, and younger hbl. andesite breccias and tuffs.
		Underwood	Tu	0 – 125		Dark brown to black hypersthene andesite, hbd. augite andesite mudflow breccias, some olivine basalt tuff.
21 m.y.→		*Valley Springs*	Tv	0 – 10		Fine-grained, white rhyolite tuff, chiefly water-laid.
PRE-TERTIARY		*Sierran basement*	Grd	>10,000		Leucocratic hbl. granodiorite (Cretaceous) with thin pendants of gneiss, schist, and hornfels (? Permian).

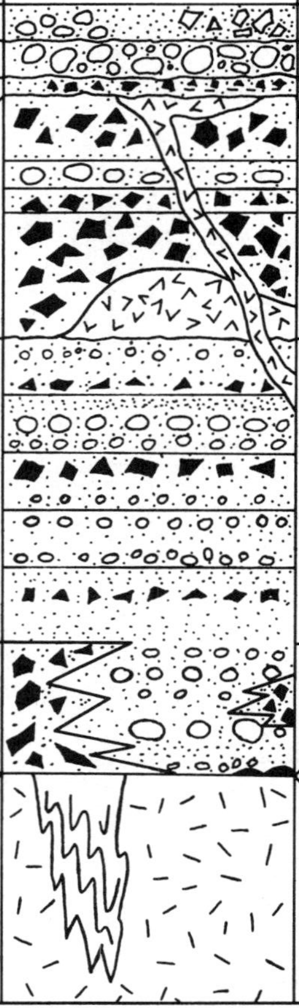

FIG. 3. Stratigraphic column of the Tertiary rocks in the Mt. Reba area, Alpine County, California.

Valley Springs Formation

The oldest Tertiary formation in this area is a white, poorly bedded, fine-grained rhyolite vitric tuff. It is exposed in three small areas: in a gulley on the south side of State Highway 4, slightly west of Lake Alpine Lodge; on the Mt. Reba road, close to its junction with State Highway 4; and on the south slope of Mt. Reba ridge below the fire lookout, where it crops out as a small horse. The tuff is scarcely 10 to 15 ft. (3-4.5 m) thick. Just off the map north of Bear Valley Inn, the tuff grades into a thin sequence of laminated claystones deposited in a lake. These rocks have yielded a small flora composed of evergreen dicots and a few deciduous hardwoods that appear to be Early Miocene. The site, which was exposed during excavation to make the recreation lake for Bear Valley, is no longer accessible.

The tuff evidently represents the Valley Springs Formation which crops out in the foothills of the range and elsewhere in the Sierra, and is Lower Miocene (K/Ar age = 22 m.y.) as judged from radiometric evidence (Dalrymple, 1964b, p. 20, table 2).

Bear Valley Formation

This name is applied here to the dark volcanic conglomerate, sandstone, and associated andesite tuffs which form the northeast wall of Bear Valley. An excellent section is exposed along the Lake Valley Road which departs from the highway to the Mt. Reba Ski Bowl at Poison Canyon (Forest Service road [jeep only] 17E01). The formation crops out widely in the region, and extends for 5 miles (8 km) to the northeast, where it laps against the granitic terrain east of Sandy Meadow (sec. 30, T-8 N., R-19 E., Markleville quad).

The lower part is chiefly andesitic sandstone and pebble to cobble conglomerate. This part of the section is characterized by augite andesite and hypersthene hornblende andesite cobbles chiefly, as well as some clasts of basement rocks, including hornels and large granitic boulders, some of which are pegmatitic. Some of these evidently were not derived from the nearby basement terrane, but may have been transported from sites now east of the Sierra, as noted by Slemmons (1953, 1966), in the Sonora Pass region to the southeast. Distant transport seems especially likely for the exotic clasts in the mudflow breccias. As exposed in the area north of Lake Alpine, the formation is made up of alternating lenses of gravel, sandstone, and andesite tuffs that rapidly vary in thickness from a few centimeters to 2 or 3 m, to 30 m or more. The section is best exposed on the steeper slopes, because it is poorly indurated for the most part. On the gentler slopes, the rocks are mantled with loosened, weathered debris, chiefly cobbles and pebbles of andesite, which cover the formation.

In this area, the total thickness of the formation is on the order of 800 ft. (240 m). It is younger than the tuff assigned to Valley Springs Formation which it overlies, as may be seen on the road to Mt. Reba Ski Bowl. It is therefore Lower to Middle Miocene, but a closer age assignment is not now possible apart from securing a K/Ar date from the interbedded andesite tuffs. Fossil wood occurs infrequently in the section, but is of no value in making a close age assignment. It is noted, however, that the tree-rings of several different taxa are widely spaced, suggesting that optimum conditions for growth were then present.

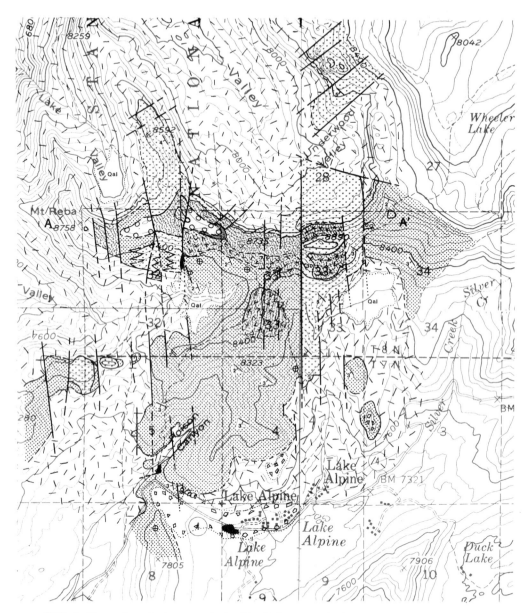

FIG. 4. General geologic map of the Mt. Reba area, Alpine County, California.

EXPLANATION

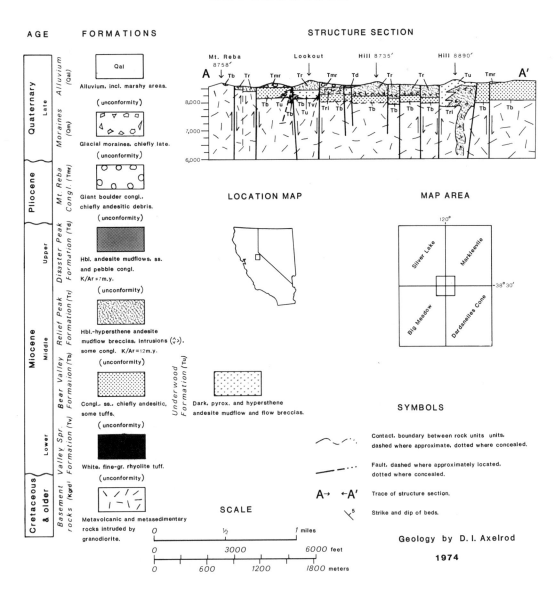

AGE

FORMATIONS

STRUCTURE SECTION

Quaternary — Late

Alluvium (Qal) — Qal
Alluvium, incl. marshy areas.

(unconformity)

Moraines (Qm) — Glacial moraines, chiefly late.

(unconformity)

Pliocene

Mt. Reba Congl. (Tmr) — Giant boulder congl., chiefly andesitic debris.

(unconformity)

Miocene — Upper

Disaster Peak Formation (Td) — Hbl. andesite mudflows, ss. and pebble congl. K/Ar=7m.y.

(unconformity)

Middle

Relief Peak Formation (Tr) — Hbl.-hypersthene andesite mudflow breccias, intrusions (◊◊), some congl. K/Ar=12m.y.

(unconformity)

Bear Valley Formation (Tb) — Congl., ss., chiefly andesitic, some tuffs.

(unconformity)

Underwood Formation (Tu) — Dark, pyrox. and hypersthene andesite mudflow and flow breccias.

Lower

Valley Spr. Formation (Tv) — White, fine-gr. rhyolite tuff.

(unconformity)

Cretaceous & older

Basement rocks (Kgrd) — Metavolcanic and metasedimentary rocks intruded by granodiorite.

Mt. Reba 8758′   Lookout   Hill 8735′   Hill 8890′

A ↓ Tb Tr   Tmr   Tr ↓   Tmr   Td   ↓ Tr   Tr   ↓ Tu   Tmr   A′

8,000

Tb Tu Tb/Tv/ Tri Tb   Tb   Tb Tb Tb Tri   Tb   Tb
Tb Tu

7,000

6,000

LOCATION MAP

MAP AREA

120°

Silver Lake   Markleeville

-38° 30′

Big Meadow   Dardanelles Cone

SYMBOLS

Contact, boundary between rock units units, dashed where approximate, dotted where concealed.

Fault, dashed where approximately located, dotted where concealed.

A→ ←A′   Trace of structure section.

↘⁵   Strike and dip of beds.

Geology by D. I. Axelrod

1974

SCALE

0   ½   1 miles

0   3000   6000 feet

0   600   1200   1800 meters

Underwood Mudflow Breccia

This name is applied to the hypersthene and augite mudflow breccia which crops out prominently on the ridges that form the western and eastern walls of Underwood Valley and extends farther east for some distance, where it also overlies the granitic basement. It also occurs on the highway to Mt. Reba Ski Bowl, and underlies the ridge that reaches eastward from Mt. Reba (see Fig. 4). The section is prominent because the dark brown to black mudflows contrast markedly with the light gray granitic basement on which it rests, and with the gray to pinkish andesites that overlie it. This dark-colored andesite unit rests on the granodiorite basement in the northern part of the area, but to the south, as in the middle of the upper part of Horse Valley and also on the road to Mt. Reba Ski Bowl, it interfingers with the Bear Valley Formation. The maximum thickness of the formation is on the order of 200 ft. (60 m), and it quickly thins out to the south in the Bear Valley Formation. The breccias are massive and poorly bedded. The highway to the ski area cuts a pipe in the formation in which clasts of olivine basalt vary from 1 cm to 5 or 6 cm long. Exotic clasts of augite andesite and hornblende andesite are present, but chiefly in the lowest part of the section. Exotic blocks of Valley Springs Formation are also in the basal beds in the formation exposed on the road to Mt. Reba Ski Bowl. The formation evidently accumulated on the low slopes of a broad valley in which the volcaniclastic sediments of the contemporaneous Bear Valley Formation were also deposited. The formation is of the same age as the Bear Valley Formation, or Lower to Middle Miocene.

Relief Peak Formation

A major unconformity cuts across the top of the dark-colored Underwood mudflow breccia and the associated Bear Valley Formation. Faulting and uplift of small blocks resulted in erosion and partial stripping of much of these units in local areas. The sequence of hornblende hypersthene andesite that accumulated on this beveled surface is chiefly pink to light gray in color, and in the field contrasts markedly with the other units. These andesites, which are rich in hornblende, are represented chiefly by mudflow breccias, brecciated flows, and intrusions. The basal part of the formation, as exposed on the saddle in section 33, 1,700 ft. (520 m) southeast of the fossil locality, is represented by a brecciated intrusion which grades upward into autobrecciated mudflows that are especially well exposed along the crest of the main ridge east of the locality. There are only a few thin associated conglomerates in that part of the section, but they increase to the west toward Mt. Reba. These are largely reworked from the autobrecciated flows. The highest part of the section exposed on the ridge 150 m east of the plant locality is a black hornblende andesite flow that weathers white to cream or light gray. A similar rock forms an intrusion on the ridge crest 3,280 ft. (1,000 m) southeast of the plant locality, and may well be the feeder for the flow.

These rocks, with a total thickness of about 320 ft. (100 m) seem in all respects similar to the Relief Peak Formation whose type area is in the Sonora Pass region 12-15 miles (19-24 km) southeast, where it exceeds 3,000 ft. (1,000 m) in thickness. Judging from my examination of both sections, the much thinner section in the Mt. Reba area appears to correspond to the upper, higher parts of the Relief Peak Formation. This inference is consistent with a K/Ar date of the hornblende andesite flow on the ridge just east of the plant locality which gives an age of 11.8 m.y. (written communication, M. Silberman, U.S. Geol. Surv., 1974).

Disaster Peak Formation

Following accumulation of the Relief Peak Formation, there was renewed faulting and erosion. A new unit accumulated on its beveled surface, but subsequent erosion has removed it from the entire area except for two small patches on the ridge east of Mt. Reba. This formation is composed of gray to brown hornblende andesite mudflow breccia and associated andesitic sandstone. The beds that yield the fossil plants are fine- to coarse-grained andesitic sandstones interbedded with thin mudflow breccias. The following section is exposed at the site which is at the head of the road that leads down to Underwood Valley (Plate 1, fig. 2).

	Approximate thickness	
Mt. Reba conglomerate (see below)		
(uncomformity)	feet	meters
concealed by slope wash	5	1.5
6. *Hornblende andesite mudflow breccia*: light gray clasts mostly 2-6 in. (50-150 mm) long, set in a gray matrix of fine microvesicular hornblende andesite tuff	3	0.9
5. *Mudflow breccia*: polymictic, clasts up to 5-6 in. (127-150 mm), but mostly 1-3 in. (25-75 mm). Chiefly hornblende andesite and hornblende pyroxene andesite.	4.5	1.3
4. *Sandstone*: andesitic to buff, cross-bedded, with rolled and twisted leaves and branchlets	2.0	0.6
3. *Andesite mudflow breccia*: polymictic, with gray andesitic matrix, clasts from 1 in. (25 mm) to 3-4 in. (57-100 mm) in diameter, chiefly hornblende andesite and rare pyroxene andesite	3.0	0.9
2. *Sandstone*: andesitic, yellow to buff, poorly bedded, poorly indurated, with twisted and rolled fossil leaves and branchlets	1.5	0.4
1. *Conglomerate*: andesitic cobbles chiefly, with thin sandstone lenses, largely concealed by slope wash	10.0	3.0
Total thickness	29+ ft.	8.6 m

(unconformity)
Pink hornblende andesite mudflow breccia
    of Relief Peak Formation.

Clasts of the light gray, microvesicular hornblende andesite mudflow breccia (unit 6) yield a radiometric age of 7 m.y., as established by Geochron, Inc. (1972); USGS (1976); and California State University, San Diego (1976).

The lower and middle parts of the coarse- to medium-grained plant-bearing sandstones contain rolled and twisted leaves and branchlets that lie at different angles to the bedding. They clearly have been transported by a moving sand body or turbidity flow.

By contrast, the upper part of the plant-bearing sandstones within an inch of the overlying mudflows are made up of crudely bedded, finer sand, and the rare leaves are flat and parallel the bedding plane. Leaves of cattail (*Typha*), which regularly inhabits the margins of ponds in shallow water, were found only in the upper part of these sandstones, thus supporting the nature of the environment adduced for this part of the section.

The plant-bearing beds rest unconformably on the light pink to gray hypersthene-hornblende andesite mudflow breccias of the Relief Peak Formation. However, a few meters down the road into Underwood Valley the plant-bearing beds are faulted against the coarse volcanic conglomerate of the Bear Valley Formation. This is the result of faulting and removal of the Relief Peak mudflows in this local area following their extravasation. As detailed below, the Disaster Peak Formation is overlain unconformably by the Mt. Reba Conglomerate on the slopes above and west of the flora.

In view of its small area, which is isolated on a ridge crest, the identity of the plant-bearing sequence cannot be established readily. The rocks are chiefly volcaniclastic hornblende andesites; but in the Sierra Nevada, mudflow breccias of this composition are of several ages, as Durrell has ably shown (1959, 1966). Nonetheless, examination of the type area of the Disaster Peak Formation, situated 14 miles (22 km) southeast, indicates that those rocks are not distinguishable from the beds at Mt. Reba. In particular, the gray, microvesicular hornblende andesite mudflow breccias seem to be inseparable. The type Disaster Peak Formation is composed in its source area chiefly of hornblende andesite mudflow breccias and flow breccias over 1,000 ft. (305 m) thick (Slemmons, 1966). From its type area at Disaster Peak, Slemmons (1966) traced the formation 15 miles (24 km) southward to the high peaks of Castle Rock and Flange Rock (alt. 9,500 ft.; 2,896 m). Thus, the formation occupies highlands in all three areas: near Mt. Reba, Disaster Peak, and Flange Rock-Castle Rock. Since the formation accumulated on a surface of low relief, its present areas have been isolated by subsequent uplift and erosion which carved the broad upland valleys and deep river canyons that now separate them.

The monomictic andesite mudflow breccia at the plant locality has a radiometric age of 7 m.y., and implies a similar age for the Disaster Peak Formation, which rests unconformably on the Stanislaus Formation, now dated at 9.5 m.y. and considered by Noble et al. (1974) to be a group. The Disaster Peak is a correlative of the Mehrten Formation (Slemmons, 1966), composed chiefly of hornblende andesite detritus carried to the foothill belt (Piper et al., 1939). As restricted by Slemmons (1966), the Mehrten overlies the Stanislaus Formation, as is apparent along State Highway 108 south of Knight's Ferry. In this area the Mehrten has yielded vertebrates (Vanderhoof, 1933) and plants (Axelrod, 1944b) of Hemphillian age.

The constructional plain that extended from the summit section down to the lowest piedmont belt was built up by the Disaster Peak-Mehrten formations well into Hemphillian time. The youngest part of the Mehrten in the foothill belt is 4-5 m.y., as indicated by the Oakdale and Turlock Lake faunas and by the Turlock Lake flora, discussed in the following chapter. Since then, the formations have been elevated, faulted, and deeply eroded to sculpture the present relief. In this regard, it is recalled that the crystalline basement on which these and the older volcanics rest, and which is exposed over a wide area as a peneplaned surface, is largely the pre-volcanic terrain. Locally it has been deeply incised by valleys of the Mountain Valley and Canyon stages of erosion, the ages of which have been disputed (see Uplift of the Sierra Nevada, below).

Mt. Reba Conglomerate

This name is given here to the local conglomerate that covers the Disaster Peak Formation on the ridge east of Mt. Reba, because of its significance in terms of the topographic history of the Sierra Nevada. The formation is a poorly lithified boulder conglomerate about 80 ft. (24 m) thick. It is composed primarily of large to gigantic, rounded to subrounded volcanic boulders, some of which are 5 ft. (1.5 m) long. They obviously were deposited by a very large river, possibly one ancestral to the present Mokelumne which now flows in the gorge 4,000 ft. (1,220 m) below Mt. Reba (elev. 8,758 ft.; 2,670 m). The clasts are mostly andesites of varied composition, and chiefly represent debris from the underlying formations in the immediate area. Clasts derived from the crystalline basement are present, but are quite rare; some of them may well be reworked from the underlying volcanic conglomerate that lies on the basement nearby.

The contact of the Mt. Reba Conglomerate with the underlying Disaster Peak Formation that yields the fossil flora is an unconformity. The rapid change in sedimentary regime from the alternating small-clast mudflow breccia and andesitic sandstone of the Disaster Peak Formation to the giant-boulder Mt. Reba Conglomerate implies greatly increased transporting power that reflects a steepened gradient, resulting in erosion of the underlying volcanic rocks in this area. As a consequence, the Mt. Reba Conglomerate rests not only on the plant-bearing beds; on the ridge half a mile west of the flora, it lies on the volcanic conglomerate of the Relief Peak Formation. Evidence for an unconformity is also seen in the presence of conspicuous platy andesitic boulders in the conglomerate that are not in the underlying beds. In addition, clasts of the distinctive microvesicular hornblende andesite that make up the uppermost flow breccia (unit 6) above the flora are reworked into the overlying conglomerate, indicating erosion of the underlying section.

The age of the Mt. Reba Conglomerate is not known, apart from the fact that it lies unconformably on the andesite breccia of the Disaster Peak Formation, which is dated at 7 m.y. However, indirect evidence presented below (see "Uplift of the Sierra Nevada" below) leads to the inference that it probably is about 5 m.y. Following deposition, the conglomerate was then faulted against the Relief Peak andesites. The record of any younger Tertiary volcanism or sedimentation is absent in this area.

Glacial Till

Glacial tills of Later Pleistocene age are spread widely in the nearby valleys at altitudes well below the fossil site. No attempt was made to discriminate between them. It is also apparent that Underwood Valley and other similar valleys in the nearby area below the present ridge crests supported small glaciers tributary to the main glacial stream in the Mokelumne gorge.

Topographic Setting

There is no evidence in this area for steep relief of 2,000-2,500 ft. (610-762 m) on the granitic-metamorphic basement as described by Slemmons (1953, 1966) for the Sonora Pass region 15-20 miles (24-32 km) southeast. Furthermore, the basement in the Mt. Reba area was not carved into rolling hills with 1,000-1,500 ft. (305-457 m) of relief as reconstructed by Wilshire (1956, 1957) for the Sierran summit area near Ebbetts Pass 10-12 miles (16-19 km) east. Only Mokelumne Peak (alt. 9,332 ft.; 2,844 m), situated across the Mokelumne River gorge 5.5 miles (8.8 km) northwest of the fossil locality, may have stood as a small isolated hill above the general peneplaned basement. There is relief on the crystalline basement in this area, but most of it is the result of later faulting. At the time of deposition of the earliest sedimentary-volcanic unit, relief on the basement evidently did not exceed 500 ft. in this area. It appears that the region was one of broad valleys separated by low ridges (see Plate 2), and was then buried by the successive volcanoclastic units. Each successive Tertiary formation is separated by a period of faulting, uplift, and erosion, during which time the area was reduced to a relatively even surface prior to the accumulation of the next unit.

Inasmuch as the Disaster Peak Formation in the Mt. Reba area is exposed only on a ridge crest in a local area, data available for reconstructing the local setting are indeed meager. Field evidence suggests that it was deposited in a broad valley carved into the underlying andesite sequence. It may be inferred that the valley walls near the plant locality were relatively low. In the first place, since Disaster Peak is separated from the underlying formations by unconformities, and considerable time (4-5 m.y.) is involved, much erosion probably occurred. Secondly, since the climate was one of well distributed rainfall and mild temperature, conditions were conducive to intense chemical weathering, and hence the production of low slopes. The valley evidently was oriented generally to the southwest, thus giving the nearby vegetation a northerly exposure and also one facing the sea, and hence well suited to plants of mesic requirements. This is consistent with the abundance of the remains of fossil Douglas fir (*Pseudotsuga*), canyon live oak (*Quercus*), and tanbark oak (*Lithocarpus*) in the deposit, for the living species closely related to them regularly inhabit moist valleys.

In summary, the Mt. Reba flora is preserved in andesitic sediments associated with mudflow breccias (K/Ar age 7 m.y.) that seem to represent the Disaster Peak Formation of the high Sierra to the southeast and the correlative Mehrten Formation in the western foothill belt. The area was characterized by very low relief. The fossil plants accumulated in a broad valley that drained southwesterly, exposed to storm tracks. The Sierran crest 12 miles (21 km) east had only low relief, and was surmounted by small, local eruptive centers and plugs.

COMPOSITION OF THE MT. REBA FLORA

The Mt. Reba flora is small, for only 14 taxa have been identified in the collection of 1,024 specimens. Others were present, as shown by a few unidentifiable leaf scraps that were found in the sandstones. As now known, the flora is made up of 7 conifers; 6 dicots, of which 2 are evergreens; and 1 perennial herbaceous monocot, the cattail (*Typha*).

Systematic List of Species

Pinaceae
 Abies concoloroides Brown
 Pinus cf. *prelambertiana* Axelrod
 Pinus cf. *sturgisii* Cockerell
 Pseudotsuga sonomensis Dorf
Cupressaceae
 Cupressus mokelumnensis Axelrod
 Juniperus sp. Axelrod
Taxodiaceae
 Sequoiadendron chaneyii Axelrod
Typhaceae
 Typha lesquereuxii Cockerell

Salicaceae
 Salix boisiensis Smith
 Salix hesperia (Knowlton) Condit
 Salix wildcatensis Axelrod

Fagaceae
 Lithocarpus klamathensis
  (MacGinitie) Axelrod
 Quercus hannibalii Dorf
Ulmaceae
 Ulmus affinis Lesquereux

The remains of *Pseudotsuga* and *Cupressus* account for 80.3% of the specimens examined (Table 2). The next most abundant plants are *Quercus* (12.6%) and *Litho-carpus* (4.7%). These 4 taxa, which make up 97.6% of the collection, inhabited moist sites, as judged from the present requirements of their nearest modern analogues. Hence they were in a favorable position to contribute abundantly to the accumulating record.

TABLE 2

Numerical Representation of Specimens
in the Mt. Reba Flora*

Fossil species	*Number of specimens*	*Percentage of flora*
Cupressus mokelumnensis leafy twigs, 578 cones, 6	584	56.8%
Pseudotsuga sonomensis (leafy twigs)	241	23.5
Quercus hannibalii	129	12.6
Lithocarpus klamathensis	48	4.7
Typha lesquereuxii	6	0.6
Salix wildcatensis	4	0.4
Abies concoloroides cone scale, 1 needles, 2	3	0.3
Sequoiadendron chaneyii (leafy twigs)	2	0.2
Salix hesperia	2	0.2
Ulmus affinis	1	0.1
Pinus cf. *prelambertiana* (broken fascicle)	1	0.1
Pinus cf. *sturgisii* (broken needle)	1	0.1
Juniperus sp. (leafy twig)	1	0.1
Salix boisiensis	1	0.1
Totals	1,024	99.08%

*Leaves, unless otherwise noted.

In this regard, it is desirable to note here that some of the *Pseudotsuga* branchlets appear to represent young growth. A similar relation is implied by some of the terminal scale leaves on the twigs of *Cupressus*. Furthermore, a number of the leaves of *Lithocarpus* also appear to represent relatively young, not fully grown specimens. The evidence leads to the tenuous suggestion that the flora may have been buried in the volcaniclastic sediments during the spring or early summer.

The remains of the other 4 conifers are exceedingly rare. Each pine is represented by only one small fragment, each of which is preserved on opposite sides of the same slab. One has the width of those produced by *P. ponderosa*. The other appears to represent the basal part of an incomplete fascicle of *P. lambertiana*. Their rarity indicates that pines were not at the immediate site. Further, since the specimens were transported in a mudflow, they evidently were derived from trees upstream, probably from an area where mixed conifer forest was dominant. This agrees with the representation of *Abies* (2 needles, 1 cone scale) and *Sequoiadendron* (3 small branchlets). Their rarity implies that they also probably lived upstream from the site of accumulation to which they were transported, or at least they were not common in the immediate area. Taken together with the preference of the modern descendants of the dominant taxa in the flora for cool sites where there is ample moisture, the extreme rarity of species that represent mixed conifer forest implies a setting below the lower margin of that forest, in a moist, well watered valley. The site of plant accumulation was dominated by a *Pseudotsuga-Cupressus* forest, with broadleaved evergreen associates which no doubt formed a woodland on well-drained sunnier slopes nearby. Most of the fossils have close relatives that still live in the nearby region to the west, in the lower part of the range. Only 2 species in the flora are not now native to California: *Ulmus* inhabits the eastern United States, and *Cupressus* has its nearest relative in eastern China, both areas with ample summer rainfall.

In the case of so small a flora, the reader may raise the question as to why the list was not supplemented by a pollen analysis. The reason is simply that a palynological study cannot provide reliable evidence to reconstruct local vegetation, or afford a basis for estimating the elevation of the flora. Fossil pollen, even if preserved in the coarse fluvio-lacustrine Mt. Reba beds, would have taxa carried from a number of other sites: the lowlands to the west; sites along the Sierran crest to the east; and vegetation zones in Nevada farther east. Clearly, such evidence would be very misleading in terms of vegetation and climate that might be reconstructed for the Mt. Reba site. In this regard, the regional derivation of well-sampled fossil pollen floras is clearly apparent from the Miocene Coatzacoaleos flora, as illustrated by Graham (1976, fig. 1). Furthermore, to provide reliable paleoecologic evidence, taxa must be identified in terms of similar living species. Inasmuch as most palynological studies rely on genera (or families) only, local paleoecologic conditions cannot be reconstructed within precise limits. In this regard, several genera in the Mt. Reba flora (*Abies, Pinus, Populus, Quercus, Salix*) have pollen that palynologists maintain usually cannot be discriminated in terms of living species. Since the species of these genera range from sea level to timberline, recognition of the genera obviously cannot provide sound evidence of composition, climate, or elevation. Furthermore, 3 other taxa in the flora (*Cupressus, Juniperus, Sequoiadendron*) are regularly grouped by palynologists as T-C-T, or grains of Taxodiaceae-Cupressaceae-

Taxaceae. Members of these families also range from sea level (e.g., *Chamaecyparis*, *Sequoia, Torreya, Cupressus*) to timberline (*Juniperus*) in California. Inasmuch as a palynological study cannot provide unequivocal paleoecological evidence regarding the Mt. Reba site, it was not undertaken.

PALEOECOLOGY

Vegetation

The general nature of vegetation in the area can be reconstructed by reference to the modern occurrences of species most similar to the fossils, as 6 of the 7 conifers are similar to those now in the central to northern Sierra Nevada and adjacent areas to the north. Douglas fir (*Pseudotsuga*) does not range much south of Yosemite, and only occurs with Sierra redwood (*Sequoiadendron*) in the Placer County Grove east of Foresthill and in the Tuolumne Grove (rare) near Yosemite. Yellow pine (*Pinus ponderosa*), sugar pine (*P. lambertiana*), and white fir (*Abies concolor*) grow with them in both areas. Tanbark oak (*Lithocarpus*) occurs in shrub form (var. *echinoides*) in the Placer County Grove of Sierra redwoods, but the tree is in Blodgett Forest 10 miles southwest in a rich Sierran mixed conifer forest. Canyon live oak (*Quercus chrysolepis*) is also in Blodgett Forest, and also occurs with *Sequoiadendron* in the southern Sierra Nevada, as in the Tule River basin and elsewhere.

Cupressus, the most abundant plant in the flora, has its nearest living relative in eastern Asia, where it is in the mixed mesophytic forest (Wang, 1961). The foliage was at first thought to represent *Chamaecyparis*, but the discovery of a few cones that are persistent on the stems indicated its real identity. The fossil resembles *Chamaecyparis torulosa* Endlicher, a species which has been considered to represent *Chamaecyparis* and has also been regarded as intermediate between *Cupressus* and *Chamaecyparis* (Dallimore and Jackson, 1966). In terms of the California flora, *Cupressus macnabiana* has branchlets that are generally in one plane, though the habit is not as marked as in the fossil, which has broad flat sprays of *Chamaecyparis*-like branchlets. Furthermore, the cones of *C. macnabiana* are much larger and are prominently horned. *C. macnabiana* inhabits somewhat drier sites in the lower part of the Sierran mixed conifer forest. It lives close to pure stands of *Pseudotsuga*, and is associated also with broadleaved sclerophyll vegetation composed of *Quercus chrysolepis* and *Lithocarpus densiflorus*, both of which have close allies in the Mt. Reba flora. In addition, *Pinus ponderosa* and *P. lambertiana* are among its associates in the Sierra Nevada, *Abies concolor* occurs a short distance upslope, and nearby riparian sites include the willows that have equivalents in the flora, as well as cattail (*Typha*). Judging from the abundant representation of *Cupressus mokelumnensis* in the flora, and from the ecological occurrence of its nearest relative in China, the fossil species probably formed part of the mesic *Pseudotsuga* forest. This agrees with its mode of occurrence in the rocks. The fossil foliage and larger twigs are typically curled and rolled in the sediment, which represents a mudflow deposit. Inasmuch as it moved down a valley, it would have chiefly picked up foliage from woody plants that lived along its path down a major drainage, not the structures of plants inhabiting drier slopes. The moist site inferred for the fossil cypress, as based on present ecology of its nearest relative, is thus consistent with its abundance and its mode of occurrence.

The 4 most abundant plants, *Pseudotsuga, Cupressus, Quercus,* and *Lithocarpus*— which account for over 97% of the sample of 1,000-odd fossil specimens (Table 2)—no doubt also lived close to the site of plant accumulation. Living species similar to 3 of the dominants are associated today (Table 3). In regions where they occur together, they regularly contribute to a *Pseudotsuga* forest and also dominate the bordering broad-leaved sclerophyll vegetation. In the Sierra Nevada, from the Sonora-Yosemite region northward to Mt. Shasta, the shady, cool, north-facing slopes of canyons at lower elevations (2,000-3,000 ft.; 610-914 m) are frequently dominated by pure stands of *Pseudotsuga,* associated with abundant *Quercus chrysolepis* along the well-drained valley walls. Judging from the rarity of the other conifers (*Abies, Pinus, Sequoiaden-dron*) in the flora, and from the absence of some that might be expected (e.g., fossil equivalents of *Torreya californica, Calocedrus decurrens*), it is inferred that mixed conifer forest lived upstream from the site of deposition.

This supposition is consistent with the distribution of forest types in the region today. *Pseudotsuga* forms relatively pure stands only at low elevations in moist, mild-winter climate in the lowest part of the Sierran forest zone not far removed from sclerophyll vegetation. The rich mixed Sierran forest lives at higher, cooler, and moister levels, either upslope on canyon walls or farther east and higher up in the range. A group of representative areas in the Sierra Nevada where these relations have been observed is

TABLE 3

Habit of Mt. Reba Species,
as Judged from Similar Living Plants

Fossil species	*Similar living species*
Trees	
Abies concoloroides	*A. concolor*
Pinus cf. *prelambertiana*	*P. lambertiana*
Pinus cf. *sturgisii*	*P. ponderosa*
Pseudotsuga sonomensis	*P. menziesii*
Cupressus mokelumnensis	*C. funebris*
Juniperus sp.	*J. occidentalis*
Sequoiadendron chaneyii	*S. giganteum*
Lithocarpus klamathensis	*L. densiflora*
Quercus hannibalii	*Q. chrysolepis*
Ulmus affinis	*U. americana*
Shrubs	
Salix boisiensis	*S. nuttallii*
Salix hesperia	*S. lasiandra*
Salix wildcatensis	*S. lasiolepis*
Herb	
Typha lesquereuxii	*T. latifolia*

presented in Table 4. The areas listed are all characterized by pure or nearly pure Douglas fir stands which are adjacent to other forest types. Upslope from the canyon bottoms and deep shaded ravines where *Pseudotsuga* dominates are subtypes in which it is a codominant with *Pinus ponderosa*, or with *Abies concolor, P. ponderosa,* and *P. lambertiana. P. ponderosa* forms pure stands on the drier, sunnier ridge crests or on south slopes. Taxa of these forest subtypes are sufficiently removed from the river border where the *Pseudotsuga* stands occur so that their structures would only rarely enter an accumulating record, if at all. The subtypes occur upstream at cooler and moister levels, and from there a few structures might be washed downstream, occasionally to enter an accumulating deposit in the *Pseudotsuga* zone.

Plate 3, Figure 1, is a view across Sutter Creek, 4 miles (6.4 km) west of Volcano,

TABLE 4
Areas of Occurrence of Pure *Pseudotsuga*
Stands in Sierra Nevada*

Quadrangle (1:125,000)	Locality	Elevation
Sonora	NE 1/4. E of Tuolumne, on North Fork of Tuolumne R.	2,500-3,000 ft.; 762-914 m
Jackson	NE 1/4. 2 mi. (3.2 km) N of Railroad Flat; 3 mi (4.8 km) W of Volcano	1,500-2,500 ft.; 457-762 m
Bigtrees	NW 1/4. Junction of Tiger Creek and Mokelumne R.; junction of Panther Creek and Mokelumne R.	2,500-3,000 ft.; 762-914 m
Placerville	NW 1/4. Hornblende Mts.; E of Placerville	1,500-2,500 ft.; 475-762 m
Smartsville	Near San Juan; near Nevada City	1,000-2,500 ft.; 305-762 m
Downieville	SW 1/4. Below Indian Valley; Indian Creek	2,000-2,500 ft.; 610-762 m
Bidwell Bar	West margin, middle. North fork of Feather River; Big Bend Mt.	1,000-1,500 ft.; 305-457 m
Redding	NW 1/4. Near Lamoine	1,500-2,000 ft.; 457-610 m
Dunsmuir	Sacramento R., below Castella	1,500-2,000 ft.; 457-610 m

*Data from vegetation type maps, California (now Pacific Southwest) Forest and Range Experiment Station, Berkeley, Calif.

Amador County, near 1,600 feet (488 m), where pure stands of *Pseudotsuga* are associated with large (some up to 3 ft. [1 m] d.b.h.), abundant *Quercus chrysolepis*. Madrone (*Arbutus menziesii*), maple (*Acer macrophyllum*), and California laurel (*Umbellularia californica*) are scattered on the slopes. The streambank is lined with maple (*A. macrophyllum*), alder (*Alnus rhombifolia*), willow (*Salix lasiolepis, S. lasiandra*), and ash (*Fraxinus oregona*). The higher slopes in the view have *Pinus ponderosa, P. lambertiana, Calocedrus decurrens*, and *Quercus kelloggii*, and *P. ponderosa* occurs also on the dry, steep, south-facing slopes behind the observer. It is emphasized that pure stands of *Pseudotsuga* occur only in narrow, deep canyons like the valley of Sutter Creek in this area. Where the valley widens to the east and west, local climate is sunnier and warmer, and hence effectively drier. There *Pseudotsuga* is quickly replaced by *Pinus ponderosa, P. lambertiana, Calocedrus decurrens, Quercus kelloggii*, and their usual associates, or is a codominant with them in a mixed forest, or is absent.

Plate 3, Figure 2, is a view into the canyon of Indian Creek northwest of Camptonville, from an elevation of 2,400 feet (732 m). It shows a pure *Pseudotsuga* forest adjacent to a dense sclerophyll woodland of canyon live oak (*Quercus chrysolepis*), tanbark oak (*Lithocarpus densiflorus*) and madrone (*Arbutus menziesii*). They are regular members of the Douglas fir forest, forming the subcanopy with maple, dogwood, and other plants. The broadleaved sclerophyll vegetation favors the warmer, south-facing slopes, whereas Douglas fir forest blankets the cooler, north-facing slopes. Higher up on the walls of the canyon along State Highway 49 toward Camptonville, the mixed Sierran forest dominates at altitudes above 2,700 feet (823 m), as seen on the upper part of the ridge to the left in this view. Among the conifers that form the canopy of the mixed forest are *Abies concolor, Pinus ponderosa, P. lambertiana, Calocedrus decurrens*, and *Pseudotsuga menziesii*. Frequent understory trees include *Arbutus menziesii, Acer macrophyllum, Quercus kelloggii, Cornus nuttallii*, and *Lithocarpus densiflorus*. *Quercus chrysolepis* is also present, but rarer. *Taxus breviflora* is scattered in the forest in moister sites. Associated shrubs include *Arctostaphylos patula, Ceanothus integerrimus, Corylus californicus, Prunus emarginata*, and many others. From the relations observed here, it is evident that vegetation at middle levels along Indian Creek finds analogy with the fossil flora. Obviously, if the Mt. Reba site had been situated in a rich mixed conifer forest, species representing it would be expected to have a larger representation in the collection. This is clear from the evidence presented above, which shows that the fossil plants were *transported to* the site of accumulation in a mudflow. Geologic and paleoecologic evidence thus indicate that mixed conifer forest lived in cooler sites farther upstream, though this does not preclude the occurrence of scattered forest stringers reaching down to the drainageway from bordering slopes.

Douglas fir forest ranges northward from the Sierra Nevada into the Klamath Mountain region, and thence southward in the Coast Ranges and northward into Oregon and Washington. It frequents mesic sites inland from the coast redwood (*Sequoia*) forest in California, and the cedar-hemlock (*Thuja-Tsuga*) forest in the region farther north. In these areas, broadleaved sclerophyll vegetation frequently borders Douglas fir forest on adjacent slopes, a community with which it shares a number of species. Among the taxa that commonly form the subcanopy of the *Pseudotsuga* forest are *Acer macrophyllum, Arbutus menziesii, Castanopsis chrysophylla, Lithocarpus densiflora, Quercus chrysolepsis*, and *Umbellularia californica*. In the Mt. Shasta region, Douglas fir dominates the lower part of the mixed conifer forest down to 1,500 feet (457 m). Locally, the northerly-facing, cooler slopes have pure stands of *Pseudotsuga*.

This area is of interest because *Chamaecyparis* occurs here, and it may be considered an ecologic equivalent of the chamaecyparoid *Cupressus* that dominates the Mt. Reba flora. This analogy is supported by its abundance, which indicates a valley bottom and stream-border habitat, much like that of *Chamaecyparis*. Port Orford cedar (*Chamaecyparis*) occurs along streambanks in the mixed conifer forest, reaching down to 2,000 feet (610 m) on the Sacramento River at Castle Crags State Park. Its lower limit appears to be determined chiefly by the high summer temperatures and a high evaporation rate which could be unfavorable for it, even though the trees are confined to the riverbank, much like alder. In this connection, Sudworth (1908, p. 175) notes that it lives chiefly in areas of moderate temperature and high humidity, as on the coastal slopes of southern Oregon and northwestern California. He states that it is sensitive to rapid changes in temperature and humidity (which occur in the interior), and suffers from prolonged drought, especially after rapid growth. On this basis, its present limited and discontinuous distribution over the interior may be due chiefly to the difficulty of establishing seedlings or young plants during hot, dry summers. When I visited Castle Crags State Park on Aug. 6, 1972, temperature at 2 p.m. was 110° F (43.4° C) and humidity very low.

Among the commoner woody plants in the mixed conifer forest along the Sacramento River in this area are the following:

Canopy trees
  Abies concolor
  Calocedrus decurrens
  Pinus lambertiana
  Pinus ponderosa
  Pseudotsuga menziesii (dominant)
Subcanopy
  Acer macrophyllum
  Cornus nuttallii
  Lithocarpus densiflorus (rare)
  Quercus chrysolepis
  Quercus garryana (drier sites)
  Quercus kelloggii
  Taxus breviflora (moist sites)
Shrub layer
  Acer circinatum
  Amelanchier alnifolia
  Ceanothus integerrimus

Cornus californica
Corylus californica
Crataegus douglasii
Prunus emarginata
Prunus demissa
Rhamnus californica
Rosa californica
Schmaltzia (*Rhus*) *trilobata*
Symphoricarpus albus
Toxicodendron (*Rhus*) *diversiloba*
Riparian-border
  Alnus rhombifolia
  Chamaecyparis lawsoniana
  Fraxinus oregona
  Rhododendron occidentale
  Salix exigua
  Salix lasiandra
  Salix lasiolepis

The preceding relations indicate that if a rich mixed conifer forest lived at the Mt. Reba site, then *Abies*, *Pinus lambertiana*, *P. ponderosa*, *Sequoiadendron*, and other taxa (e.g., *Calodedrus decurrens*), might reasonably be expected to have a good representation in the flora. Furthermore, since broadleaved sclerophyll taxa now occur in the lower, warmer part of the Sierran mixed conifer forest, the rarity (e.g., *Abies*, *Pinus*, *Sequoiadendron*) or absence (e.g., *Calocedrus*, *Torreya*) of other conifers and their associates suggests that mixed conifer forest probably attained optimum development at moister, higher altitudes in the nearby region. This seemingly explains the poor representation of *Abies* (1 cone scale, 2 needles), *Pinus* (1 fragmented needle of each species on the same fist-sized specimen), and *Sequoiadendron* (3 very small branchlets) in the fossil collection, though they may have lived on nearby slopes removed from the site of plant accumulation, or possibly were only very rare members of the immediate flora. Regardless of their exact source, they nonetheless indicate that a rich mixed conifer forest of *Abies, Pinus, Pseudotsuga, Sequoiadendron,* and probably others lived in mesic, cooler sites removed from the immediate area of plant deposition, dominated by *Cupressus, Lithocarpus, Pseudotsuga,* and *Quercus.*

Finally, reference is made to the single branchlet of *Juniperus* that is similar to those of the living *J. occidentalis,* according to Frank C. Vasek, who has studied the specimen. This species ranges from the upper part of the mixed conifer forest to higher levels where it is a member of the subalpine zone, frequently inhabiting drier, south-facing rocky slopes. As pointed out elsewhere (Axelrod, 1976a), Tertiary species allied to conifers that are now restricted chiefly to the subalpine forest in the Sierra Nevada reached down to lower altitudes in the Miocene and Pliocene. Under a climate with a longer precipitation season, summer rain, and more equable temperature, they then contributed to a mixed conifer forest that was richer in composition than any that has survived. This is shown by the occurrences of fossil species closely allied to *Abies magnifica, Picea breweriana, Pinus monticola, Alnus tenuifolia, Populus tremuloides, Sorbus scopulina,* and others that reached well down into the lower mixed conifer forest zone into Pliocene time. As summer drought increased with the emergence of montane mediterranean climate, they evidently were confined chiefly to the upper part of that belt and to the subalpine zone in the Sierra. On this basis, *Juniperus* probably occurred chiefly on drier ridges and slopes provided by outcrops of volcanic rocks. To judge from its rarity, it probably was situated upstream, in the area from which the other rare conifers (*Abies, Pinus, Sequoiadendron*) contributed to the Mt. Reba record.

We may conclude from the evidence provided by the relative abundance of the fossils, by their mode of occurrence, and from inferences based on the distribution of closely related taxa in modern forests, that the Mt. Reba area was dominated by a *Pseudotsuga-Cupressus* forest that included species of *Lithocarpus* and *Quercus* in the understory. The latter two contributed also to broadleaved sclerophyll vegetation on nearby warmer slopes. A rich mixed conifer forest composed of *Abies, Pinus, Pseudotsuga,* and *Sequoiadendron* evidently attained optimum development farther upstream, and in that area dry sites provided by outcrops of volcanic mudflows probably supported *Juniperus. Salix* inhabited streambanks and moist seepages and flats in the area. The rare record (1 leaf) of *Ulmus* in the flora implies that it probably lived upstream, in the mixed conifer forest where precipitation was higher.

Climate

The climate under which the flora lived may be inferred in part from the requirements of modern forests allied to the fossil communities which have been reconstructed for the area. As we have seen, a *Pseudotsuga-Cupressus* forest covered cooler valley slopes, evergreen sclerophyll forest blanketed warmer slopes, and a rich mixed conifer forest lived in cooler, moister sites at slightly higher levels to the east.

Total precipitation was generally similar to that which occurs today in areas where *Pseudotsuga* is associated with *Lithocarpus* and *Quercus chrysolepis*, and where mixed conifer forest with *Abies, Pinus, Pseudotsuga, Sequoiadendron*, and others cover nearby slopes. Data from meteorological stations in the Sierran forests that show relationship to the fossil flora suggest that the Mt. Reba flora probably received 35-40 in. (762-889 mm) at a minimum. Some of it must have been distributed in the warm season, for the flora includes *Ulmus* and a species of *Cupressus* whose nearest descendants are in regions with summer rain. Furthermore, fossil floras of similar age in the lowlands to the west, such as the Oakdale (Axelrod, 1944b) and Mulholland (Axelrod, 1944a), have taxa (e.g., *Karwinskia, Nyssa, Robinia, Sapindus*) that indicate summer rain. Since they were less abundant as compared with earlier in the Neogene, minimum rainfall is implied for their existence. This may have totaled 4-5 in. (101-127 mm) during the three warmest months, though the amount would depend on its effectiveness, as determined chiefly by summer temperatures which were more moderate than those at present.

The thermal conditions under which the flora lived can be estimated in two ways. The first method uses the general thermal requirements of the major vegetation zones represented in the flora. Figure 5 plots the temperature for a number of representative stations from Yosemite north to Mt. Shasta village. Since the fossil flora resembles vegetation near the upper margin of *Pseudotsuga* forest and broadleaved sclerophyll vegetation, at a minimum the mean annual temperature was probably near 55-57°F (12.8-13.9°C), or an average of 56°F (13.3°C). In the Sierra Nevada, the range of mean monthly temperature (A) is from 30 to 34°F (16.7-17.8°C) in areas where vegetation shows relationship to the Mt. Reba flora (Fig. 5). That it was lower during Hemphillian time is indicated by evidence of milder winters, as inferred from the presence in the Mulholland, Oakdale, and other floras of taxa (e.g., *Karwinskia, Persea, Sabal, Sapindus*) that are now in Mexico. There also are species in the Mulholland flora that now have their nearest descendants (e.g., *Lyonothamnus floribundus, Malosma laurina, Ceaonthus spinosus, Quercus tomentella*) confined to coastal or insular southern California, an area of very equable climate (M 65-70). Furthermore, the absence at this time (7 m.y.) of major icecaps would result in milder winters, as would the occurrence of warmer seas at this latitude. Not only were winters milder, summer temperatures were not as high as those of today. In this regard, mixed conifer forest and broadleaved evergreen sclerophyll forest occur in the inner Coast Ranges (near Angwin, Ukiah, Willits Howard Forest, St. Helena) in areas where mean monthly range of temperature is fully 4-6°F (2.2-3.3°C) lower (i.e., 26-28°F; 14.4-13.6°C) than that in the Sierran community, and the range of temperature may be only 20-25°F (11.0-13.9°C) in mountains nearer the coastal strip. The difference between the mean January and July temperatures was therefore certainly less than 30°F (16.7°C). That it was lower than that now in the Sierra is consistent also with the poorer development of broadleaved sclerophyll vegetation there as compared with the Coast Ranges. This evidently is the result of colder winters which freeze the evergreens, and heavy snows which regularly break their branches. Both factors appear responsible for the wide restriction of the community there during the Quaternary ice ages (Axelrod, 1976b). The data suggest that

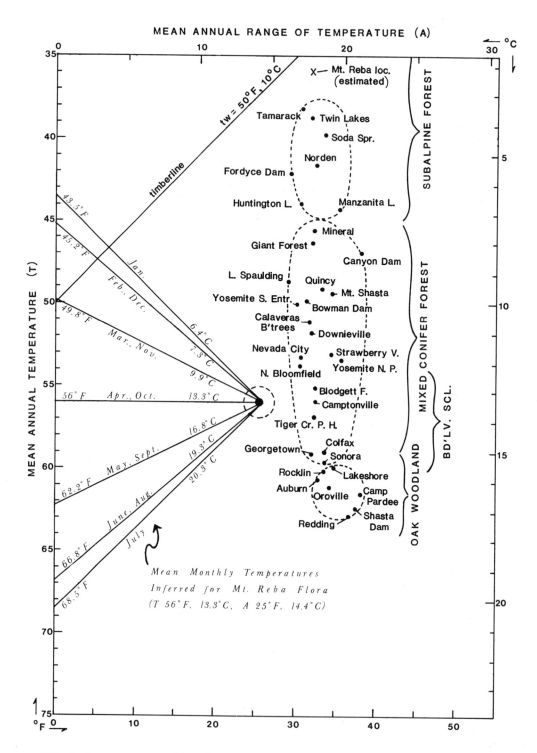

FIG. 5. Temperature conditions in the Sierran forests today, and inferred thermal regime of the Mt. Reba flora.

when compared with present conditions in the Sierran forests, the difference between the mean July and January temperature probably was near 24-26°F (13.3-14.4°C). On this basis, the Mt. Reba flora probably lived under a mean annual temperature (*T*) of about 56°F (13.3°C) and a mean annual range of temperature (*A*) of approximately 25°F (13.9°C).

These estimated temperatures yield the mean monthly temperatures shown in Figure 5, as determined from a method devised by Bailey (1960, 1964) and discussed elsewhere (Axelrod and Bailey, 1968, 1976). For the conditions postulated, warmth of climate (*W*), or effective temperature, was about 56.8°F (13.8°C), indicating that there were 176 days of the year with mean temperature above 56.8°F (13.8°C). The temperateness, or equability, rating for the synthetic thermal environment which has been reconstructed is *M* 60, and there were only about 3 percent of the hours (e.g., 261 hours) of the year with frost.

The second method of estimating the thermal conditions of the Mt. Reba flora utilizes temperatures at meteorological stations where vegetation similar to the flora lives today. Camptonville (elev. 2,800 ft.; 853 m) is in the lower part of the mixed conifer forest dominated by *Pseudotsuga*, with associates of *Abies concolor, Calocedrus decurrens, Pinus ponderosa*, and *P. lambertiana*, and a subcanopy of *Acer macrophyllum, Arbutus menziesii, Cornus nuttallii, Lithocarpus densiflorus, Quercus chrysolepis, Q. kelloggii*, and their usual shrubby associates. As noted earlier, in the valley of Indian Creek 4 miles (6.4 km) northeast, and reaching down to the North Yuba River (Plate 3, Fig. 2), the north slope is dominated by Douglas fir forest with a subcanopy of *Q. chrysolepis* and *Lithocarpus*, which both contribute to sclerophyll woodland on warmer exposures. This zone reaches from near 2,600 ft. (792 m) down to 2,200 ft. (670 m) on the north Yuba River just west of Indian Valley, and 6 miles from Camptonville. Climatic conditions at Camptonville are not greatly different from those indicated by modern vegetation in Indian Valley. However, the latter area has a milder summer climate, for the vegetation zones are on north-facing slopes at lower altitudes (Plate 2, Fig. 2), whereas Camptonville is on a rolling, broad interfluve fully exposed to the summer sun. Camptonville has mild winters, for in no month is the average minimum temperature below freezing; the Mt. Reba flora is estimated to have had cooler summers and milder winters (Fig. 6).

Farther upstream on the north Yuba River, Downieville (elev. 2,895 ft.; 882 m) is situated in the lower mixed conifer forest dominated by *Pseudotsuga*. *Q. chrysolepis* is a common associate and also dominates the warmer, south-facing canyon walls. However, this area shows less relationship to the Mt. Reba flora than the vegetation of Indian Valley, 10 miles (16 km) west and 500 ft. (152 m) lower. That area marks the inland (eastward) distribution of both tanbark oak (*Lithocarpus*) and madrone (*Arbutus*), which are abundant in the Camptonville area and on the slopes of Indian Creek. As judged from the climatic records at Downieville, it has a climate too cold in winter for *Arbutus* and *Lithocarpus*. Although mean January temperature at Downieville is 37.5°F (3.1°C), there are five months in which average minimum temperatures are below freezing (Fig. 7). Clearly, the fossil flora had warmer winters, and as noted above there is evidence to indicate frost was not as frequent.

Georgetown (elev. 2,640 ft.; 705 m) is situated on a broad rolling interfluve between the North and Middle Forks of the American River, 45 miles (72.4 km) south of Downieville. The area is relatively sunny and exposed, and is therefore dominated by a *Pinus ponderosa* forest. However, nearby sheltered and moister canyons and valleys have Douglas fir forest with *Q. chrysolepis*, as in the valley of Georgetown Creek 3 miles

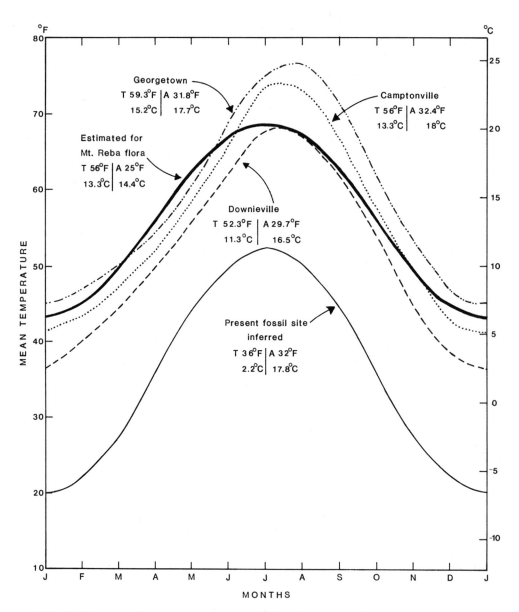

FIG. 6. Mean monthly temperatures at or near sites where vegetation shows relationship to the Mt. Reba flora, compared with temperatures inferred for the fossil flora. Approximate thermal conditions at Mt. Reba today are also indicated.

(4.8 km) west of Georgetown, at 1,800-2,000 ft. (549-610 m). *Lithocarpus* is not far to the northeast where precipitation is greater. It forms a valley forest with *Pseudotsuga* and *Q. chrysolepis* in the drainage of Otter Creek, from 3,000 ft. (914 m) down to 2,500 ft. (762 m). This community dominates north-facing slopes and is surrounded by mixed conifer forest. As compared with vegetation represented by the Mt. Reba flora, Georgetown has

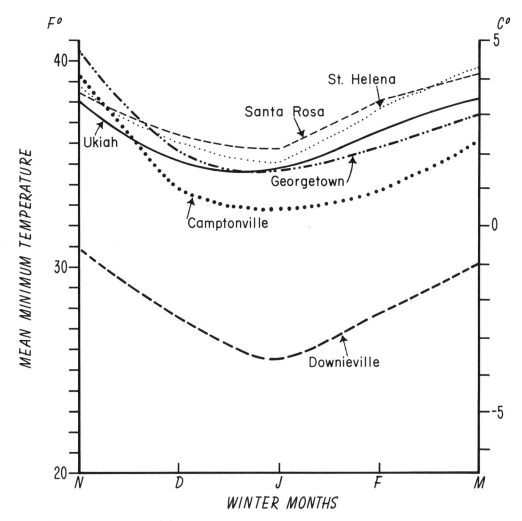

FIG. 7. Present mean minimum temperatures used to estimate winter conditions for the Mt. Reba flora.

a warmer, sunnier climate (Fig. 6). However, it and the nearby area have very mild winters (Fig. 7), and would readily support the Mt. Reba community—as is evident from relations in the valley of Otter Creek. Slightly higher mean winter temperatures are recorded in the Coast Ranges (Ukiah and St. Helena, Fig. 7), where *Arbutus, Lithocarpus*, and *Q. chrysolepis* are regularly associated with *Pseudotsuga* and find more optimum development. Georgetown has no month in which the average minimum temperature is below freezing: the mean daily minimum January temperature is 34.6°F (1.4°C).

Blodgett Forest (elev. 4,350 ft.; 1,326 m), situated several miles east of Georgetown, has a rich mixed conifer forest with *Lithocarpus* and *Q. chrysolepis* in the subcanopy. Other sclerophylls in the area include *Arbutus* as well as large trees of *Chrysolepis*

(= *Castanopsis*) *chrysophylla* which is otherwise known only from the coastal region. The area is 10 miles southwest of the northernmost *Sequoiadendron* grove, which is at an altitude of 5,200 ft. (1,585 m). The grove is in a west-facing canyon, sheltered from cold winter blasts and open to moderating maritime influence in both summer and winter. Temperatures are more moderate in the grove than at Blue Canyon, which is situated on a ridge crest 10 miles (16 km) north, also at 5,200 feet (1,585 m). Comparative temperature data for Blodgett Forest and Blue Canyon are shown in Table 5. The North Grove of Sierra redwoods probably has intermediate conditions. The occurrence there of the shrub form of *Lithocarpus* (var. *echinoides*) and the absence of *Arbutus, Chrysolepis chrysophylla,* and *Quercus chrysolepis* indicate a colder climate than at Blodgett Forest. Since the mean daily minimum January temperature at Blue Canyon is 28.5° F (-1.9°C), it probably is close to 32° F (0°C) in the *Sequoiadendron* grove, an environment too cold for some of the equivalents of the Mt. Reba flora.

TABLE 5

Comparative Thermal Regimes at Two Stations
in the Central Sierra Nevada

		Mean temperature	
Stations	*January*	*July*	*Annual*
Blodgett Forest*			
elev. 4,350 ft.	38.8°F	71.5°F	55.2°F
(1,326 m)	(3.8°C)	(21.9°C)	(12.9°C)
Blue Canyon			
elev. 5,280 ft.	37.1°F	68.0°F	52.5°F
(1,609 m)	(2.8°C)	(20.0°C)	(11.4°C)

*10-year record, kindly furnished by Herbert C. Sampert, School of Forestry and Conservation, University of California, Berkeley.

The two methods of estimating temperature give similar results. Note that in Figure 6 the actual monthly temperatures at modern stations show a displacement to the right, especially in spring, when compared with the idealized curve drawn from the mean annual temperature (T) and the mean annual range of temperature (A). That is, in spring the sine wave indicates a rise that is premature for the actual temperatures, which show a lag by a week or more. Further, the actual recorded temperatures do not behave the same on the rise in spring as on the decrease in autumn. This reflects the fact that in spring, solar heat is going into the ground, and so the air is relatively "robbed" of warmth. But in autumn, heat is being released from the ground, and hence the air is enriched by heat. Furthermore, air circulation in summer is more active in the vertical than in winter, owing to the strength of summer sunshine. This tends to bring to the instrument shelter influences of the relatively stable atmosphere above, and especially so if the site is well elevated. Summer maxima thus tend to be delayed somewhat, as if the instruments were near the ocean. By contrast, in winter the frequency of calm, cold nights is sufficiently great to bring close together in time the longest night of the year and the coldest time of the year. These general phenomena have been considered by Leighly (1938).

The marked changes in thermal conditions adduced for the Mt. Reba area since the flora lived there are indicated by comparison with present-day temperatures at nearby stations (Twin Lakes, Tamarack) in the subalpine forest zone at levels close to 8,000 ft. (2,438 m). Conditions estimated for the present Mt. Reba site (elev. 8,650 ft.; 2.637 m), which is about 600 ft. (183 m) below regional timberline (elev. 9,300 ft.; 2,835 m) are shown in Figure 5. Mean annual temperature at the fossil site has been lowered approximately 20°F (11.1°C) from about 56°F (13.3°C) to 36°F (2.2°C). Clearly, the area has been uplifted into a colder, more extreme climate, with a shorter growing season.

Elevation

A measure of the approximate amount of uplift in the Mt. Reba area is provided by an analysis of the elevation of modern vegetation similar to that in the fossil flora. However, the climatic conditions (precipitation, temperature) that control the upper and lower limits of forest zones today are not the same as those of Pliocene or Miocene times. Hence, the data provided by the elevation of present-day communities similar to the fossil flora must be modified in terms of the nature of past climate, insofar as it can be inferred.

In reconstructing late Neogene environment, it is recalled that the Coast Ranges were not yet elevated appreciably. As a result, a near-maritime influence extended much farther inland than it does at present, bringing temperatures more nearly like those now in the Coast Ranges to the forest zones on the west slope of the Sierra. Further, a warmer sea occupied the outer coastal strip, flooding much of the San Joaquin Valley and lapping against the west front of the southern half of the Sierra. Higher marine temperature produced a climate of milder winters, and also was a source for summer rainfall, which no longer occurs in the region. As a result, there were in the region a number of fossil taxa whose nearest descendants now occur in coastal and insular southern California (*Lyonothamnus, Ceanothus spinosus, Rhus laurina*), or in the Southwest (*Robina, Sapindus, Quercus* sp.), or in Mexico (*Persea, Karwinskia*).

During the Neogene, rainfall was more evenly distributed over the region because terrain was lower, and hence precipitation was not concentrated as at present on the windward slopes of the Coast Ranges and the Sierra at the expense of interior valleys situated in their shadows. Hence the present high rainfall now in the central Sierra Nevada, where many living relatives of the flora occur, does not provide a measure of that which may have been required to support the fossil flora. In fact, all of the species range into areas with considerably lower precipitation.

Under the Pliocene climate of milder summer and winter temperature and some summer rainfall, drought stress was reduced as compared with that now affecting the living derivative species, which live under a full mediterranean climate with a long, hot, dry summer season. As a result, taxa were able to live at lower levels than do their modern descendants, with ecotypes of aspen, fir, spruce, Sierra redwood, and others reaching down to the live-oak woodland zone (Axelrod, 1941, 1976a, 1976b).

With respect to the elevation of the major vegetation zones, the preceding discussion of physical conditions indicates that the area was dominated by Douglas fir forest with broadleaved sclerophyll vegetation on warmer slopes, and with mixed conifer forest in the nearby area, but restricted primarily to somewhat cooler, higher levels. Similar relations occur today in the Sierra Nevada at elevations up to about 3,000 ft. (914 m) in the region west of the flora, and at 2,000 ft. (610 m) in the region farther north, where precipitation is higher (Table 3). Among the upper sites where vegetation shows relationship to the fossil flora, we have noted the valley of Indian Creek east of Camptonville at levels near 2,500-2,700 ft. (762-823 m) (Plate 3, Fig. 2), and the valleys of Otter Creek and Canyon Creek east of Georgetown at an elevation of from 2,500 ft. (762 m) to about 3,000 ft. (914 m).

There are, of course, local areas at higher levels in the Sierra where *Pseudotsuga* forest and broadleaved sclerophyll vegetation are close to mixed conifer forest. For instance, a large patch of *Q. chrysolepis* and *Pseudotsuga* reaches up to 4,700 ft. (1,433 m) on the south-facing wall of Clavey Creek, Sonora Quadrangle (Critchfield, 1971). However, it is the exposure of the area that increases warmth locally and favors the extralimital occurrence there. That this patch may be a relict of the Xerothermic Period (8,000-4,000 years ago) is quite possible, to judge from relations in the Sierra and elsewhere (Axelrod, 1966, p. 45). In any event, relief like that at Clavey Creek certainly could not have been in the Mt. Reba area, for it was the upper part of a broad, rather featureless volcanic plain sliced by drainages with relatively low valley walls. There are local enclaves of *Pseudotsuga* forest and broadleaved sclerophyll vegetation in the mixed conifer forest where warmer sites favor their occurrence. However, such a setting is not consistent with the nature of the fossil flora. As discussed earlier, it indicates some transport, implying that the site was well up in the Douglas fir-*Cupressus*-sclerophyll forest belt, and not within the bordering mixed conifer forest.

Inasmuch as the Mt. Reba site evidently was not higher than the lower edge of mixed conifer forest, a *maximum* elevation near 3,000 ft. (914 m) seems indicated on the basis of *modern relations* of vegetation and climate in the Sierra Nevada. However, Pliocene climate was more temperate, hence the climatic (and therefore vegetation) belts must have been at a somewhat lower level. Whereas the range of mean monthly temperature (*A*) is from 30 to 34°F (16.7-17.8°C) in the Sierra today (Fig. 5), in the late Neogene it was somewhat lower and probably more nearly like that now in the Coast Ranges, where mixed conifer forest lives above Douglas fir forest and broadleaved sclerophyll woodland. In the region from near Mt. Helena (Sonoma-Lake counties border) and northward, the transition into mixed conifer forest is frequently near 2,000-2,500 ft. (610-762 m). Climatic data for stations in the Coast Ranges near areas where broadleaved sclerophyll vegetation and Douglas fir forest mingle with mixed conifer forest (e.g., Angwin, Upper Lake Ranger Station, Ukiah, Willits Howard Ranger Station) shows that the range of temperature is between 26 and 28°F (14.4-15.6°C) today. It is somewhat less in the bordering hills, where the broadleaved sclerophyll forest also occurs, for temperatures there are more moderate than in the stations, which are chiefly in valleys. Thus a range of about 25°F (13.9°C) between the mean January and July temperature seems likely.

The range of annual temperature in the lowlands can be inferred also from the frost requirements of avocado (*Persea*), as illustrated in Figure 8. *Persea* reproduces naturally in areas where there is ample summer rain and winters are largely free from frost. In California, it is grown commercially along the coastal strip from near Santa Barbara to San Diego, with irrigation providing water during the dry summer season. Climate at

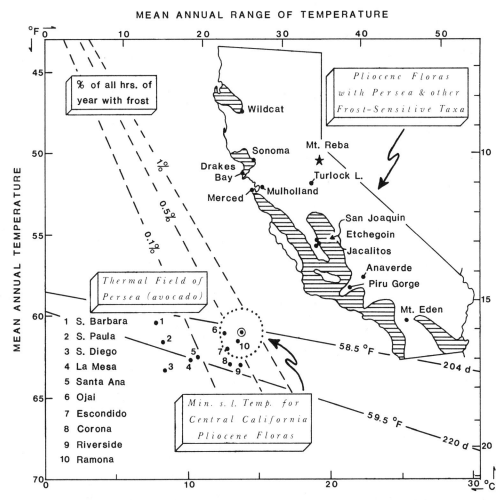

MEAN ANNUAL RANGE OF TEMPERATURE

FIG. 8. Sea-level temperatures in central California during the Middle Hemphillian (7 m.y.), as inferred from frost requirements of avocado (*Persea*) and other frost-sensitive taxa in these floras. Approximate area of marine embayment shown by hachured area.

the inner limit of cultivation, as marked by several stations in Figure 8, has a frost frequency of about 0.5 percent (43 hours) of the year. *Persea* was a common tree in the Miocene and Pliocene of California. It occurs at all fossil sites shown in Figure 8 except the Mulholland, which nonetheless has comparable frost-sensitive taxa, notably species allied to *Ceanothus spinosus*, *Lyonothamnus floribundus*, and *Malosma laurina* that are now in coastal southern California (Axelrod, 1944a), and which occur there bordering avocado groves, or are in even milder, insular sites as shown by *Lyonothamnus* and *Dendromecon* (cf. *harfordii*).

In addition, *Persea* is associated in floras farther south (e.g., Anaverde, Piru Gorge) with palms, and in others (Anaverde, Mt. Eden) with frost-sensitive thorn scrub taxa (e.g., *Acacia*, *Eysenhardtia*, *Ficus*). The data suggest that at the base of the Sierra west of

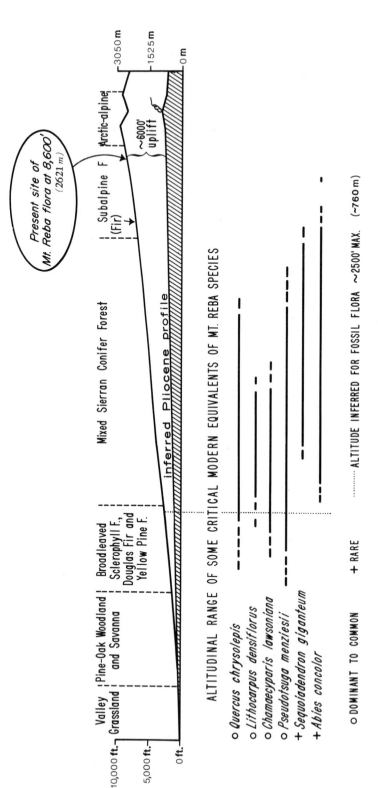

FIG. 9. Topographic-vegetation profile across the central Sierra Nevada during Middle Hemphillian time (7 m.y.). The Mt. Reba flora was near 2,500 ft. (762 m), in the upper Douglas fir-broadleaved sclerophyll forest, whereas the site is now 6,000 ft. (1,828 m) higher, in the upper subalpine forest zone.

Mt. Reba the mean range of temperature probably was not much more than 25°F (13.9°C); otherwise frost frequency (see Fig. 8) would be too great for the taxa that are recorded there and in nearby parts of central California. This is similar to the estimate reached on the basis of the range of temperature indicated by vegetation in the Coast Ranges allied to that in the fossil flora. The data also imply that mean temperature was near 62°F (16.7°C). With a range of 25°F (13.9°C), warmth of climate was W 59°F (15°C), or 212 days with mean temperature above that level. This also corresponds to the warmth of climate now at the inner limit of *Persea* cultivation in southern California. This appears to be a fair estimate of minimum temperature at sea level at the base of the Sierra, inasmuch as *Persea* is rare there and increases in abundance toward the coastal strip and southward.

The preceding inferences imply a difference in mean annual temperature (T) between Oakdale (T 62°F; 16.7°C) and Mt. Reba (T 56°F; 13.3°C) floras of about 6°F (3.3°C). Using a terrestrial lapse rate of -3.0°F/1,000 ft., or 183 m/-C° (see Axelrod and Bailey, 1976), the Mt. Reba site would have had an altitude near 2,000 ft. (333 ft. × 6°F = 1,996 ft.), or 610 m. Realizing that this is only an estimate, an altitude between 2,000 and 2,500 ft. (610-762 m) seems likely. As noted above, the rarity of taxa representing mixed conifer forest in the sample also suggests that altitude probably did not exceed 3,000 feet (914 m), as judged from *modern* conditions. It probably was not below 2,000 ft. (610 m), because oak woodland taxa are not recorded in the flora. On this basis, the general level of the Sierran summit region to the east probably did not exceed 3,000-3,500 ft. (914-1,067 m), apart from a few small, scattered volcanic centers reported by Curtis (1951, 1954), Wilshire (1956, 1957) and Slemmons (1953, 1966). If we accept a maximum elevation near 2,500 ft. (762 m) for the Mt. Reba flora, then uplift at the fossil site amounts to about 6,000 ft. (1,829 m) since the Middle Pliocene (7 m.y.), as sketched in Figure 9. These results, reached from a paleoclimatic analysis of plant evidence, are similar to the estimate by Hay (1976) based on plate tectonics, which gives an uplift of 1,800 m (5,900 ft.) in 4.5 m.y. (also see "Uplift of the Sierra Nevada" below); my estimate was during the first draft of this paper, five years earlier.

AGE

The Mt. Reba flora has a modern aspect. Its taxa are very similar to living species, most of which live in the nearby region at lower elevations. Floras of this nature did not appear in central California until the Hemphillian time (Axelrod, 1944d). This is demonstrated by the pronounced differences between the Table Mountain (Condit, 1944) and Oakdale (Axelrod, 1944b) floras of the lower slopes of the central Sierra Nevada, or between the Neroly (Condit, 1938) and the Mulholland (Axelrod, 1944a) floras of the Mt. Diablo region to the west. In both areas fully 30 to 40 percent of the species in the older floras no longer have close analogues in the region, whereas less than 20 percent of the taxa in the younger floras have their nearest descendants in distant regions (Table 6); the somewhat higher percentage for the small Oakdale flora probably results from an inadequate sample.

Percentage data of this sort have age significance only in a local area, not in more distant regions (Axelrod, 1957); climatic change did not induce modifications in composition at the same rate everywhere, as some have assumed. Floras of relatively modern aspect appeared much earlier in the interior than in the coastal regions. For instance, the

TABLE 6

Percentage of Exotic Woody Taxa
in Some Floras of Hemphillian Age
in California

Flora	No. of species	Exotic Taxa	
Mt. Reba	13	*Cupressus mokelumnensis, Ulmus affinis*	15%
Mulholland	41	*Acer bolanderi, Berchemia multinervis, Mahonia marginata, Nyssa elaenoides, Populus parcedentata, Populus washoensis, Quercus bockeei, Sapindus oklahomensis*	19
Piru Gorge	21	*Acer arida, Persea coalingensis, Sabal mohavensis*	14
Mt. Eden	47	*Arbutus prexalapensis, Dodonaea californica, Eysenhardtia pliocenica, Ficus edenensis, Juglans beaumontii, Persea coalingensis, Populus sonorensis, Sapindus oklahomensis*	17
Oakdale	16	*Populus parcedentata, Robinia californica, Sapindus oklahomensis, Mahonia oakdalensis*	25

Eocene Bull Run flora (40 m.y.) of northeastern Nevada has 35 genera, 25 (70 percent) of which still live within 150 miles of the site. By contrast, only 15 percent of the genera in the essentially contemporaneous Comstock flora of western Oregon are still in that region.

The data in Table 6 also contrast sharply with evidence provided by the few small Hemphillian floras now known east of the Sierra Nevada. The Verdi (Axelrod, 1958) has 16 woody plants, but only 2 (*Populus subwashoensis, P. payettensis*) have descendants that are now in distant regions, which gives the flora 12 percent exotic taxa. Scarcely 20 percent of the species in the combined late Barstovian-early Clarendonian Aldrich Station, Fallon, and Chloropagus floras (*Betula, Populus, Sophora, Ulmus*) of west-central Nevada (Axelrod, 1956) have their nearest descendants in distant areas, yet the contemporaneous Table Mountain and Neroly floras, west of the Sierra, have 35 to 40 percent exotic taxa. In this regard, the Pliocene Sonoma flora from the coastal strip near Santa Rosa, California, has 35 species, of which 6, distributed in *Castanea, Ilex, Mahonia, Persea, Ulmus*, and *Trapa* (17 percent), are no longer in the region (Axelrod, 1944c). They persisted longer there because mild maritime climate damped the stress of high evaporation in summer, which was typical in the interior and unfavorable for the persistence of relicts which required summer rain. Likewise, the similar percentage of exotics in the Hemphillian Mulholland flora, as compared with the Fallon, Aldrich Station, and other Nevada floras, reflects its near-coastal position. This is emphasized by the occurrence at Mulholland of several taxa (*Lyonothamnus, Quercus, Ceanothus, Malosma, Schmaltzia*) whose closest allies are in insular or coastal southern California, as well as in the southwestern United States (Table 6). The Piru Gorge flora (Axelrod, 1950b) of Hemphillian age also has taxa that have their nearest relatives in the coastal area (e.g., *Prunus, Ceanothus*), as does the contemporaneous Mount Eden flora of interior southern California, with species of *Pinus, Ceanothus, Juglans, Malosma* (Axelrod, 1950a). The trend to less equable conditions in these areas since Hemphillian time resulted in part from secular change, and in part from major topographic changes. Uplift of the mountain ranges between the fossil localities and the coastal strip has resulted in increased heat in summer and greater cold in winter over the interior, a trend that commenced earlier in the interior, east of the Sierran axis.

The Middle Hemphillian age of the Mt. Reba flora, as suggested by its composition, agrees with evidence provided by three radiometric dates secured from hornblende in the monomictic andesite mudflow breccia a few feet above the plant-bearing beds (see Disaster Peak Formation, under "Geology" above) which yield an average age of 7 m.y. Its age in relation to other floras in the adjacent region is indicated in Table 1.

UPLIFT OF THE SIERRA NEVADA

Paleoecologic and geologic relations of the Mt. Reba flora provide critical evidence with respect to the age of the uplift of the range. As outlined in the "Geology" section of this chapter, the Mt. Reba flora is preserved in rocks on a summit ridge that represents an old erosion surface beveled across the granitic basement (on Mt. Reba), two older andesitic formations stratigraphically below the flora, and a giant boulder conglomerate that lies unconformably above it (Fig. 3). Deposition of each formation was separated by a period of faulting, uplift, and erosion prior to the accumulation of the next unit, a history paralleling that documented by Durrell (1959, 1966) in the northern Sierra. During each period of erosion, the terrain was reduced to a surface of low relief. Each succeeding Tertiary formation, which in most cases represents a blanket deposit, diminishes gradually in thickness from the summit region toward the western foothill area. Thus the west slope was primarily a constructional plain on which a new consequent drainage was established by streams as each new volcaniclastic unit accumulated. Only locally in the higher parts of the nearby Sierra do ancient crystalline highs, as Mokelumne Peak, appear to have escaped inundation by the volcanic outpourings (Piper et al., 1939). These authors and others (e.g., Lindgren, 1911; Christensen, 1966) have also considered that the Pyramid Peak range probably stood well above the volcanic field; in fact, Lindgren (1911, p. 185) regarded it as the early Tertiary drainage divide of the Sierra Nevada. My observations indicate that it is a major horst, bounded on the west by a fault that trends northwesterly through Dark Lake-Wrights Lake from the high scarp at the east edge of Silver Lake on the Carson Pass Highway 88, a fault that is clearly post-andesite in age. On the other hand, in the lower central foothill belt metamorphic rocks locally provided a series of long, low discontinuous ridges that evidently were largely buried by Miocene volcanic formations, as discussed in the next chapter.

The giant-boulder Mt. Reba Conglomerate (Plate 1) perched on Mt. Reba ridge is 4,000 feet (1,219 m) above the Mokelumne River, which flows in a deep gorge 2.5 to 3 miles (4-4.8 km) northwest (Plate 2). The river that deposited the conglomerate was large, fully equal to those now flowing in the lower part of the range. Clearly, the Mokelumne River cut its present gorge since the Mt. Reba Conglomerate was deposited. The age of the conglomerate is not precisely known, but it rests unconformably on the plant-bearing beds (7 m.y.); it may be of essentially the same age, or possibly as much as 2-3 m.y. younger (see below). In any event, it is clear that the present upland surface that forms the ridge crests of the region was beveled *after* the conglomerate was faulted, and is therefore post-7 m.y. in age. Between the ridge crest that trends eastward from Mt.

Reba and the deep canyon of the Mokelumne River, there are upland valleys carved into the crystalline basement well below the contact with the volcanic sequence, and fully 500-600 ft. (150-180 m) or more below the ridge crest. These include Underwood, Lake, and Horse valleys on the south side of the river, and their counterparts across the Mokelumne Gorge are Tanglefoot Canyon, Deadwood Canyon, and others (Plate 2). They border the river gorge and increase in size to the west, at lower elevations.

These three topographic surfaces correspond to those that have been identified by Curtis (1951), Slemmons (1953), and Wilshire (1956) in the nearby Sierra to the east and southeast, respectively. These authors refer to the Broad Valley, Mountain Valley, and Canyon stages of erosion recognized by Matthes (1930a) in the Yosemite area. Matthes thought that the Broad Valley surface was cut during the Miocene, the Mountain Valley in the Pliocene, and the Canyon during the early (and later) Pleistocene. At Mt. Reba, the Broad Valley landscape (ridge crest) was still being sculptured across granitic basement, volcanic rocks, and the conglomerate well into Hemphillian (middle Pliocene) time. The Mountain Valley and Canyon stages of erosion are therefore younger than the youngest Mehrten sediments, or on the order of 4 to 5 m.y., as indicated in the following chapter on the Turlock Lake flora.

This does not agree with the age that has been assigned to the time of uplift and carving of the Canyon Stage in the region 100 miles (160 km) southeast. In that area, Dalrymple (1963, 1964) found that in the North, Middle, and South Forks of the San Joaquin River, basalt flows dated at 3.1 and 3.5 m.y. rest on the Mountain Valley surface and spill over into the canyon below, indicating that the canyons had largely been excavated by that time. This conclusion evidently was supported by evidence in the Kern River canyon, where a basalt flow, also dated at 3.5 m.y., lies within a few hundred feet of the bottom of the 3,000-foot-deep (914 m) canyon. These data have led to the conclusion that "An average of 300 ft. (91 m) of erosion, concentrated in the canyons, has occurred since early Pliocene time, but the canyons had nearly their present depths 3.5 m.y. ago, suggesting completion of tilting by then" (Bateman and Wahrhaftig, 1966, p. 158).

The conflict in these estimates of the age of canyon cutting and the time of tilting is apparent. In similar kinds of massive plutonic rocks 100 miles (160 km) apart, erosion was 7 times more rapid in the Mokelumne than in the San Jaoquin River drainage. Further evidence of the discrepancy is found in the upper part of the San Joaquin River basin. At San Joaquin Mountain (elev. 11,000 ft.; 3,353 m) the oldest volcanic rocks are basaltic andesite flows dated at 3.5 m.y. (Dalrymple, 1963, 1964) that rest on the crystalline basement. There is general agreement that the lavas flowed west, down a broad valley that extended across the present range (Matthes, 1930b; Erwin, 1934; Huber and Rinehart, 1965a, 1965b). The flows now form the crest of the ridge on the east side of the San Joaquin River gorge, perched 2,000 ft. (609 m) above the river. Clearly, the present canyon is younger than these flows. Further evidence for the recency of the canyon cutting is seen several miles southeast of San Joaquin Mountain, where the McGee till rests on basalt dated at 2.7 m.y. The till was deposited by a glacier that flowed down a broad upland valley, prior to the excavation of the present canyon of McGee Creek which is 3,000 ft. (914 m) deep.

Pollen evidence recovered from fine-grained sediment under the basalt flows at San Joaquin Mountain indicates that a mixed conifer forest allied to the Sierra redwood (*Sequoiadendron*) forest on the west slope of the range inhabited this area in the late

Pliocene (Axelrod and Ting, 1960). This evidence has been reevaluated by Christensen (1966, p. 172) and Curry (1968), who assert that the presence of *Pinus flexilis* and *P. murrayana* in the flora indicates that the region already was at subalpine levels, and that the taxa that represent plants like those now on the west side of the range were transported from that area to the east side of the Sierra at San Joaquin Mountain. Actually, neither *P. murrayana* nor *P. flexilis* is an exclusive indicator of a subalpine environment. *P. flexilis* regularly descends into the *lower part* of the mixed conifer forest in areas where there is some summer rainfall, and also on the east side of the Sierra near Bridgeport. In the Sweetwater Mountains to the northeast, it reaches well down into the pinyon pine-juniper belt. *P. flexilis* also contributes to a mixed forest at local sites along the east front of the southern Sierra and in the San Bernardino Mountains, as does *P. murrayana*. Furthermore, in areas with more summer moisture, as in the Spring (= Charleston) Mountains of southern Nevada, and in Utah and Colorado, *P. flexilis* is a regular member of the *Abies concolor-P. ponderosa-Pseudotsuga menziesii* forest. Since there was still effective summer rainfall in California in the Pliocene, these pines cannot be considered as a reliable indicator of subalpine conditions, and hence (indirectly) of the altitude of the area—as Christensen and Curry believe.

As for the transport of the San Joaquin Mountain pollen flora largely from the west side of the range, Christensen and Curry have not taken into account the occurrence of a similar flora at Owens Gorge (Axelrod and Ting, 1960) which also represents vegetation like that at or near the present Sierra redwood groves on the west side of the range. Furthermore, Ting and Axelrod analyzed modern pollen rain (counts of 500 grains) in the area of San Joaquin Mountain from six sites ranging from the upper subalpine forest east of Minaret Summit to Agnew Pass 8 miles north at 11,500 ft. (3,505 m). The pollen rain at each site is composed chiefly of plants in the nearby area today.[1] This agrees with the results of Adam (1967, pp. 279-283), who studied the modern pollen rain along transects from the Central Valley across the Sierra Nevada through Carson Pass and Tioga Pass and into the Great Basin. If, as Christensen and Curry suppose, the San Joaquin Mountain pollen was transported to the fossil site from the western slopes of the range, which Christensen asserts (1966, p. 172) had essentially its present elevation by the close of the Pliocene, then why is a similar flora not recorded there by the pollen rain today?

The fossil pollen evidence, meager though it is, suggests a setting near 5,000 ft. (1,530 m) and under conditions not greatly unlike those now on the west slope of the Sierra near the Tuolumne and Stanislaus groves of Sierra redwood, but with more summer rain. The relations provide a basis for inferring that the present crestal area formed by the Minarets, which now reach 13,000 ft. (3,962 m) directly west of the San Joaquin Mountain flora, was then near 8,000 ft. (2,438 m). This would have been adequate for a subalpine forest of *Pinus flexilis*, which most probably extended to lower levels, and for its associates of *P. murrayana*, *P. monticola*, and others that contributed occasional grains to the accumulating record in the Sierra redwood forest in the valley at the eastern base of the Pliocene Minaret Range.

This reconstruction is consistent with evidence provided by the Deadman till which is

---

1. This example of modern pollen rain is not to be confused with the preceding discussion in the section on "Composition of the Flora." A pollen analysis of the Mt. Reba sediments could not reveal local conditions, because the sediments obviously have been transported. However, pollen recovered from sedimentary rocks of a fossil lake probably does depict the general nature of the vegetation in the drainage basin.

overlain by latites (2.7 m.y.) and underlain by andesite (3.1 m.y.) on the upper slopes of San Joaquin Mountain, and is the oldest well-dated glacial deposit in the Sierra (Curry, 1966). The till, which is interbedded with volcanics that form the crest of San Joaquin Mountain, crops out discontinuously for 10 miles (16 km) from Minaret Summit to Agnew Pass. The clasts that make up the till were derived from the Minaret Range to the west (Curry, 1966), which is now separated from the Deadman till by the 2,000-foot-deep (610 m) canyon of the San Joaquin River. This again implies that in this area the Canyon stage is younger than the Deadman till and the latites (2.7 m.y.) that overlie it on the crest of San Joaquin Mountain. It also implies that uplift to form the present stream gradients is more recent, which is consistent with the displacement shown by the volcanics and tills on the east front of the range in the nearby area, which suggest at least 4,000 ft. (1,219 m) of uplift since ~3.0 m.y. ago.

In this regard, it must be added that Curry resurrects Matthes' notion that the present graben area east of the Sierran crest, which is represented now by the present Owens Valley trough, formed a great arch in the Late Pliocene. Following the extravasation of the andesites and basalts dated at about 3.0 m.y., he infers that the Pliocene crestal area to the east collapsed, leaving the present fault trough. As pointed out earlier (Axelrod and Ting, 1960, p. 39), this notion is inconsistent with pollen evidence from sites now in the axis of the trough, notably in Owens Gorge and also in the slightly younger (2.7 m.y.) Coso Formation to the south. They show, together with the pollen flora from San Joaquin Mountain, that the region along the east front of the Sierra and down the present graben supported a mixed conifer forest that indicates only moderate altitude. The absence of subalpine forest and arctic alpine taxa dominating the beds is also consistent with only moderate altitude. The few taxa that are in the floras that range up into the subalpine zone, notably *P. murrayana*, *P. aristata-flexilis*, etc., are not sufficiently well represented to indicate that they dominated the area. Also, the recorded floras are predominantly west-side in affinity and indicate mild climate. As noted above, samples of modern pollen rain along the Sierran summit region in the subalpine zone, and in the arctic alpine zone above it, show that there are no (or few) taxa being transported across the Sierra in any abundance today, as Curry and Christensen assert. Furthermore, Curry's notion that the Pliocene treeline was probably higher than the present one (he cites no evidence) is contrary to climatic principles. Under the moister and cooler Late Pliocene climate, and with the lower range of mean monthly temperature than at present, *all* forest zones are depressed; the Pliocene subalpine zone was lower then than it is today.

The evidence for the recency of canyon cutting in this sector of the range agrees with data from the northern Sierra (Durrell, 1959, 1966). There the Warner Basalt, which appears to be contemporaneous with the Tuscan Formation, is disrupted by faulting which has displaced it fully 3,000 ft. (914 m) (Durrell, 1966). Furthermore, the Feather River Canyon has been incised 3,000 ft. (914 m) through the Warner Basalt and down into the metamorphic basement, as may be seen near Beiber in the valleys of Indian Creek and Chips Creek (T 25 N, R.6 E, on Westwood Sheet, Geol. Map of California). Even in the lower foothill belt, the Tuscan Formation, dated at 3.3 m.y. (Evernden et al., 1964), has been incised by the valley of Butte Creek fully 1,000 ft. (305 m), and by the west branch of the Feather River 1,200 ft. (366 m). This indicates that the *present canyons* in the northern Sierra are post-Tuscan and post-Warner basalt in age, and hence commenced to develop just prior to the first Sierran glacial episode, or soon thereafter.

In summary, the evidence reviewed here suggests that the canyons of the Sierra were not already excavated to nearly their present depths by the later "Pliocene" ($\sim$ 3.0 m.y.), as implied by radiometric evidence furnished by basalts in the San Joaquin and Kern river drainages. It seems more probable that they were incised after major uplift(s) which postdate the basalts. The discrepancy may well be the result of dating basalts which were contaminated as they welled up through the plutonic basement, much as some samples of the Columbia River Basalt have also provided unreliable ages (Gray and Kittleman, 1967).

The preceding data make it feasible to estimate the rate of erosion in the middle and upper parts of the Sierra Nevada during the later Cenozoic. At San Joaquin Mountain, the San Joaquin River has cut to a depth of 2,500 ft. (762 m) in 3 m.y., or at a rate of $\sim$ 800 ft. (244 m)/m.y. At Mt. Reba, the Mokelumne River canyon has been deepened fully 4,000 ft. (1,219 m). The time required is not closely established but is certainly less than 7 m.y. (the age of the Mt. Reba flora), to judge from the nature of the unconformity between the Mt. Reba Conglomerate and the Disaster Peak Formation. Using the same rate for the San Joaquin River at San Joaquin Mountain, situated 75 miles (121 km) southeast and also carved into plutonic rocks chiefly, we can make several estimates, as follows: 5 m.y. x 800 ft./m.y. = 4,000 ft. (1,219 m); 6 m.y. x 800 ft./m.y. = 4,800 ft. (1,463 m); 7 m.y. x 800 ft./m.y. = 5,600 ft. (1,707 m). The first estimate is consistent with the evidence and supports the inference of a similar rate of erosion in these nearby areas. The estimate of $\sim$ 5 m.y. also provides a clue to the probable age of the Mt. Reba Conglomerate. In addition, this time probably corresponds closely to the major uplift which inaugurated the Mountain Valley stage.

The conclusion that the major episode of uplift in the Sierra occurred well after the time of the Mt. Reba flora finds further support in current evidence in the San Joaquin Valley at the western base of the range. Grant, McCleary, and Blum (1977) show that in the area between the drainages of the Merced and Mokelumne rivers, rate of tilting has been approximately 24 ft. per mile per m.y., commencing near the close of Mehrten deposition. Since the dips of the Mehrten, Stanislaus (= Table Mountain), and Valley Springs formations are approximately the same, they conclude that significant tilting of the Sierra block did not occur between the Oligocene and Pliocene, but resulted from uplift along the east front and has been relatively uniform since 4-5 m.y. ago.

That the major period of uplift may be later is implied by data from Owens Valley (Bachman, 1978). The Waucobi Lake beds (2.3 m.y.) provide a record of the development of Owens Valley and the uplift of the White-Inyo mountains block following formation of the eastern Sierra scarp by downwarping, preceding the deposition of the lake beds, and followed by faulting and volcanism. That the Mt. Reba conglomerate may reflect this warping, which would have increased stream gradients down the western slope about 3.5 m.y. ago, seems likely.

SUMMARY

The Mt. Reba flora occurs near timberline in a volcaniclastic section similar to the Disaster Peak Formation, in the nearby high Sierra Nevada, and to the Mehrten

Formation (restricted) in the foothill belt. The plant-bearing beds rest unconformably on older andesite formations which lie unconformably on rhyolite tuff or on crystalline basement. The section is overlain by the Mt. Reba Conglomerate, deposited by a river whose ancient course is now perched on a ridge crest at 8,700 ft. (2,652 m), fully 4,000 ft. (1,219 m) above the present Mokelumne River. Periods of faulting and erosion separate each Tertiary formation.

The flora is composed chiefly of taxa allied to those now in the lower part of the Sierran forest dominated by *Pseudotsuga, Lithocarpus,* and *Quercus* (cf. *chrysolepis*), and with *Abies, Pinus,* and *Sequoiadendron* in cooler, moister sites upslope. *Lithocarpus* and *Quercus* also contributed to broadleaved sclerophyll vegetation on warmer slopes. Yearly precipitation was near 40 in. (1,016 mm), with sufficient summer rain to support species of *Cupressus* and *Ulmus* no longer in the far West. Mean monthly temperatures were more moderate than in related modern forests, with winters mild and snow of brief duration.

As judged from the modern distribution of related vegetation, and inferred paleo-temperature data, the site had an elevation near 2,500-3,000 ft. (762-914 m), implying that the area has been since uplifted about 6,000 ft. (1,828 m) to its present position near timberline.

The middle Hemphillian age of the flora is indicated by its composition, and by radiometric dates (K/Ar age = 7 m.y.) of the associated hornblende andesite.

Previous reports that the major river canyons in the central and southern Sierra Nevada were largely excavated by the end of the "Pliocene" (3.5 m.y.) appear incorrect, and evidently were based on dating contaminated basalt flows. Evidence in the Mokelumne-Stanislaus river drainage area shows that the Mountain Valley stage was cut in response to uplift following deposition of the Mt. Reba Conglomerate, which unconformably overlies the Disaster Peak Formation. The conglomerate is younger than 7 m.y., and plate tectonic evidence implies that it probably is 5 m.y., consistent with rates of erosion ($\sim$ 800 ft. or 244 m/m.y.) in the Mokelumne and upper San Joaquin River basins. The Canyon cycle commenced after deposition of the Tuscan Formation (3.3 m.y.) in the northern Sierra, and after the oldest Sierran glaciations (2.7-3.0 m.y.) in the Mammoth region (Deadman Till, McGee Till).

SYSTEMATIC DESCRIPTIONS

Family PINACEAE
Abies concoloroides Brown
(Plate 6, fig. 3)

Abies concoloroides Brown, Wash. Acad. Sci. Jour., vol. 30, p. 347; Chaney and Axelrod, Carnegie Inst. Wash. Pub. 617, p. 138; pl. 11, fig. 4, 1959 (see synonymy); Axelrod, Univ. Calif. Pub. Geol. Sci., vol. 33, p. 275; pl. 4, figs. 2-6; pl. 12, figs. 6-8; pl. 17, figs. 5, 6; pl. 25, fig. 5, 1956; Axelrod, *ibid.,* vol. 39, p. 227; pl. 42, figs. 1-3, 1962; Axelrod, *ibid.,* vol. 51, p. 105; pl. 5, figs. 4-10; pl. 7, fig. 1, 1964.
Abies concolor Wolfe (not Lindley), U.S. Geol. Surv. Prof. Paper 454-N, p. N14; pl. 1, fig. 10; pl. 6, figs. 1-3, 6, 10, 11, 1964.

The record of fir in the Mt. Reba flora is limited to 2 needles and a partly shriveled cone scale that appears to have been exposed to drying prior to burial. The fossils are similar to the smaller needles and cone scales produced by the living white fir, *Abies concolor* (Gordon and Glend.) Lindley of the western United States.

Collection: U.C. Mus. Pal., Paleobot. Ser., Mt. Reba, hypotype no. 5948; homeotype nos. 5949-5950.

Pinus cf. *prelambertiana* Axelrod

Pinus prelambertiana Axelrod, Univ. Calif. Pub. Geol. Sci. vol. 34, p. 127; pl. 18, figs. 9-12, 1958.

A broken and mangled portion of the basal part of a fascicle composed of quite thin needles seems referable to this species. It is the only record of this taxon in the flora and, as noted above, occurs on the same slab that has a broken needle which has been referred to *P.* cf. *sturgisii* Cockerell. Transport from a more distant area upstream probably accounts for the rare records of these taxa in the flora.

Collection: U.C. Mus. Pal., Paleobot. Ser., Mt. Reba, homeotype no. 5951.

Pinus cf. *sturgisii* Cockerell

Pinus sturgisii Cockerell, Amer. Jour. Sci. ser. 4, vol. 26, p. 538, text-fig. 2, 1908.

The material in the Mt. Reba flora referred to this species is only a part of a single needle from a presumed 3-needle fascicle. In thickness and keeled structure it is readily matched by needle fragments of the living western yellow pine. This specimen, as well as that referred above to *P.* cf. *prelambertiana*, probably were broken during transport to the Mt. Reba site from an area to the east, where they evidently lived. It is noteworthy that both specimens occur on opposite sides of the same slab.

Collection: U.C. Mus. Pal., Paleobot. Ser., Mt. Reba, homeotype no. 5952.

Pseudotsuga sonomensis Dorf
(Plate 4, figs. 1-3)

Pseudotsuga sonomensis Dorf, Carnegie Inst. Wash. Pub. 412, p. 72, pl. 6, figs. 2-4, 1930; Axelrod, *ibid.* 553, p. 191, pl. 36, fig. 3, 1944; Axelrod, *ibid.* p. 251, pl. 42, fig. 1, 1944; Chaney and Axelrod, *ibid.* 617, p. 143; pl. 13, figs. 13-15, 1959 (see synonymy); Axelrod, Univ. Calif. Pub. Geol. Sci., vol. 39, p. 228; pl. 42, figs. 10, 11, 1962.

Numerous branchlets, some of which are up to 13 cm long with attached needles, represent this species in the Mt. Reba flora. The needles are slender, of variable length on different specimens, and display a distinct, slightly twisted petiole and a broadly acute apex. The fossils are similar to leafy branchlets of *P. menziesii* (Mirb.) Franco, which ranges southward in the Sierra Nevada to the Yosemite region, and is distributed in the Coast Ranges from central California northward into the Pacific Northwest.

No cones were found in the plant beds, nor were any winged seeds or cone scales recovered that might represent this species.

Collection: U.C. Mus. Pal., Paleobot. Ser., Mt. Reba, hypotype nos. 5953-5955; homeotype nos. 5956-5966.

Family CUPRESSACEAE
Cupressus mokelumnensis n. sp.
(Plate 5, figs. 1, 3; plate 6, figs. 1, 3)

Description: Branchlets up to 15 cm long, strongly flattened in one plane; branchlets 1.5 mm wide, scale leaves in opposite pairs, appressed, tips broadly acute, nonglandular. Ovulate cones 10-15 mm in diameter, globose, scales without prominent umbos, the cones evidently persistent.

Discussion: This cypress is represented by numerous branchlets that are flattened in one plane, and by the casts of 6 ovulate cones. Its nearest relatives are in eastern Asia and central Mexico. The Asian species *C. funebris* Endlicher and *C. torulosa* Don are characterized by flattened, spray-like branches very similar to those of *Chamaecyparis*. *C. funebris*, which is closest to the fossil, has been considered to be *Chamaecyparis* by some authors. Dallimore and Jackson (1966) regard it as intermediate between these genera in some respects. Inasmuch as the cones of the fossil species evidently matured in the second year, not the first as in *Chamaecyparis*, it is judged to represent *Cupressus*.

C. funebris inhabits the floristic-rich hill country of western Hupeh and Szechuan, where it contributes to a mixed conifer hardwood forest that lives under a mild temperate climate with ample summer rain. *C. torulosa* also occurs in Szechuan and ranges into the western Himalayas, where it is in mixed forests at elevations ranging from 5,000 to 9,000 ft. (1,524-1,843 m). In Mexico, *C. lusitanica* var. *benthamii* also shows relationship, though it produces scale leaves that are more awl-like than the fossils', and the branchlets seem to be slenderer than the fossils'. However, the cones are much like the fossil casts.

To judge from its associates and from its nearest living allies, *C. mokelumnensis* lived in a climate of mild temperature, and with some summer rainfall. Since it is abundantly represented together with *Pseudotsuga*, as well as *Lithocarpus* and *Quercus* (*chrysolepis*), which are mesic forest indicators, it seems probable that Mokelumne cypress contributed to a rich forest.

Of the 10 purported species of *Cupressus* in California, only *C. macnabiana* shows a distant relation to *C. mokelumnensis*. It has a tendency for the branchlets to be in one plane, but this habit is not as pronounced as in the fossil or in the Mexican or living Oriental species. In addition, the cones of *C. macnabiana* are much larger and the cone scales have prominent horn-shaped umbos.

Two valid fossil *Cupressus* species are known from Neogene floras in the west: *C. preforbesii* Axelrod from the Mt. Eden flora, and *C. mohavensis* from the Tehachapi flora. They resemble living species in California and Arizona, and are quite unlike *C. mokelumnensis*. A small terminal fragment of a leafy twig in the Sonoma flora was described as *Cupressus sonomensis* (Axelrod, 1944c). It is a *Chamaecyparis*, and its glandular scale leaves indicate it is related to *C. lawsoniana*. It is here transferred to *Chamaecyparis sierrae* Condit.

Collection: U.C. Mus. Pal., Paleobot. Ser., Mt. Reba, holotype no. 5967, paratypes nos. 5968-5970, 5971-5987.

Juniperus sp. Axelrod
(Plate 6, fig. 4)

A fragmentary impression of a well-preserved juniper twig was found on my last visit to the Mt. Reba locality. The specimen measures 34 mm long, it is branched, and the individual branches are 1 mm wide. The tips of the oppositely arranged scale leaves are broadly acute. It seems unwise to designate such a fragment as the type of a species, because it cannot readily be distinguished from juniper specimens present in other floras.

The specimen was subjected to a detailed scrutiny by Prof. Frank C. Vasek, who has had considerable experience with the morphology of the western junipers (Vasek, 1966). It is his opinion that the specimen is more nearly related to the montane *J. occidentalis* Hooker than to *J. californica* Carriere of the lower, drier woodland belt. This is expectable, for it probably would have occupied drier volcanic sites afforded by the valley walls, composed of mudflow breccias chiefly. It presumably attained more optimum development at higher levels in the range, and to the east.

Collection: U.C. Mus. Pal., Paleobot. Ser., Mt. Reba, no. 5988.

Family TAXODIACEAE
Sequoiadendron chaneyii Axelrod
(Plate 4, figs. 4-6)

Sequoiadendron chaneyii Axelrod, Univ. Calif. Pub. Geol. Sci. vol. 33, p. 280; pl. 4, figs. 25-27; pl. 5, fig. 103; pl. 17, figs. 1-4; pl. 25, fig. 4, 1956; Axelrod, *ibid.*, vol. 39, p. 228, pl. 43, figs. 3-5, 1962; Axelrod, *ibid.*, vol. 51, p. 111; pl. 7, figs. 17-20, 1964.

The record of this species is based on the terminal ends of 3 very small branchlets, 15-17 mm long. The awl-shaped needles and their spiral arrangement indicate that they represent Sierra redwood. No cones were found in the plant-bearing beds. The rarity of the fossil remains of *Sequoiadendron* in the deposit suggests that they probably were carried to the site from a more distant area upstream where the trees lived.

The Mt. Reba occurrence is the second Tertiary record of *Sequoiadendron* from the Sierra Nevada. An earlier, previously unreported record is in the andesite mudflow sequence at Frenchman Dam in the northern Sierra Nevada, north of Chilcoot, collected by Cordell Durrell.

Collection: U.C. Mus. Pal., Paleobot. Ser., Mt. Reba, hypotype nos. 5989-5991.

Family TYPHACEAE
Typha lesquereuxii Cockerell
(Plate 5, fig. 2)

Typha lesquereuxii Cockerell, Amer. Mus. Nat. Hist. Bull., vol. 24, p. 79; pl. 10, fig. 46, 1908.

Several fragmentary leaf blades in the collection are characterized by numerous parallel veins and the absence of any midrib. Some of the blades are up to 25 mm wide. The largest specimen, which was on a large block that had rolled down the slope below the locality, was fully 3 ft. (90 cm) long.

Remains of cattail occur only in the uppermost part of the plant-bearing beds, within 2-3 cm of the overlying mudflow breccias. They are flat in the matrix, not curled and twisted like most of the leaves and branchlets of conifers in the lower and middle part of the plant-bearing sandstones. The cattail leaves evidently were buried in standing water, suggesting a habitat much like that in which the plant lives today.

Collection: U.C. Mus. Pal., Paleobot. Ser., Mt. Reba, hypotype no. 5992; homeotype nos. 5993, 5994.

Family SALICACEAE
Salix boisiensis Smith
(Plate 6, figs. 7, 8)

Salix boisiensis Smith, Amer. Mid. Naturalist, vol. 25, p. 498, pl. 2, fig. 3; pl. 4, fig. 8; 1941; Axelrod, Carnegie Inst. Wash. Pub. 553, p. 194; pl. 36, figs. 8-10, 1944; Chaney and Axelrod, *ibid.*, Pub. 617, p. 153, pl. 16, figs. 9, 10, 1959.

Two nearly complete, well preserved, obovate leaf impressions are referred to this species, which it closely resembles. The fossil has more secondaries than some of the illustrated fossils, but all of them can be matched by the leaf variation of the living *S.*

scouleriana Barratt, a widely distributed small tree in the western United States. Leaves of *S. sitchensis* Sanson also show relationship to the fossil.

Collection: U.C. Mus. Pal., Paleobot. Ser., Mt. Reba, hypotype nos. 5995, 5996.

Salix hesperia (Knowlton) Condit.
(Plate 6, fig. 6)

Salix hesperia (Knowlton) Condit, Carnegie Inst. Wash. Pub. 553, p. 41, pl. 4, fig. 7, 1944 (see synonymy and discussion).

Portions of two large willow leaves, each with distinctive looping secondaries and intersecondaries, are similar to leaves of *S. hesperia* (Knowlton) Condit. They resemble leaves of the living *S. lasiandra* Bentham, a widely distributed small tree along stream borders in the western United States.

Collection: U.C. Mus. Pal., Paleobot. Ser., Mt. Reba, hypotype no. 5997; homeotype nos. 5998, 5999.

Salix wildcatensis Axelrod
(Plate 6, fig. 5)

Salix wildcatensis Axelrod, Carnegie Inst. Wash. Pub. 553, p. 132, 1944 (see synonymy and discussion); Axelrod, Univ. Calif. Pub. Geol. Sci., vol. 33, p. 287, pl. 25, fig. 14, 1956.

The basal two-thirds of a well-preserved fossil leaf displays all the features of this species, which is similar to leaves of the living *S. lasiolepis* Bentham. This is a widely distributed willow in California and adjacent Baja California that has a disjunct occurrence in southeastern Arizona.

Collection: U.C. Mus. Pal., Paleobot. Ser., Mt. Reba, hypotype no. 6000.

Family FAGACEAE
Lithocarpus klamathensis (MacGinitie) Axelrod
(Plate 6, fig. 9; plate 7, figs. 6-9)

Lithocarpus klamathensis (MacGinitie) Axelrod, Carnegie Inst. Wash. Pub. 553, p. 197, pl. 37, fig. 4, 1944 (see synonymy).
Quercus simulata Knowlton. Condit, *ibid.*, p. 45, pl. 5, fig. 3, 1944.

Numerous fossil leaves in the Mt. Reba beds resemble those of tanbark oak, *L. densiflorus* (Hooker and Arnott) Rehder. This densely branched evergreen ranges rather continuously from southwest Oregon into the San Francisco Bay region, and thence discontinuously southward in the Coast Ranges to the mountains above Santa Barbara and Ojai. Its range in the central and northern Sierra is much more discontinuous, possibly because climate is less equable there.

Collection: U.C. Mus. Pal., Paleobot. Ser., Mt. Reba, hypotype nos. 6003-6005; homeotype nos. 6006-6015.

Quercus hannibalii Dorf
(Plate 7, figs. 1-5)

Quercus hannibalii Dorf, Carnegie Inst. Wash. Pub. 412, p. 86, pl. 8, fig. 9 only, 1930; Chaney and Axelrod, *ibid.* 617, p. 168; pl. 24, fig. 2; pl. 25, figs. 11-13, 1959 (see synonymy and discussion); Axelrod, Univ. Calif. Pub. Geol. Sci., vol. 51, p. 118, pl. 11, figs. 3-9; pl. 12, figs. 1-3, 9, 1964.

Leaves of this species, which are similar to those of the living *Q. chrysolepis* Liebmann, are well represented in the fossil flora. Most of them are entire. Whether this has any ecologic significance is not now known. Nonetheless, it is noted here that trees in warmer, mild-winter climates at lower altitudes in California often produce chiefly entire leaves, whereas those in the more extreme (less equable) sites tend to have leaves that are more frequently serrated. Serrate leaves also are more frequent on young trees than old ones.

Collection: U.C. Mus. Pal., Paleobot. Ser., Mt. Reba, hypotype nos. 6016-6020; homeotype nos. 6021-6034.

Family ULMACEAE
Ulmus affinis Lesquereux

Ulmus affinis Lesquereux, Harvard Mus. Comp. Zool. Mem., vol. 6, p. 16, pl. 4, fig. 4 (part 1878); Tanai and Wolfe, U.S. Geol. Surv. Prof. Paper 1026, pl. 3, figs. B, D, E, G, 1977 (see synonymy and discussion).

Portions of two leaves in the flora appear to represent an elm, and compare favorably with those of the living *U. americana* Linnaeus which is widely distributed in the eastern United States. Some uncertainty as to its modern affinities must remain, however, because the margins of the specimens are not complete and the venation there is not preserved. The species has been recorded previously as *U. californica* Lesquereux from the Table Mountain and Remington Hill floras on the western Sierran slope, as well as at several other sites in central California, as reviewed by Tanai and Wolfe (1977).

Collection: U.C. Mus. Pal., Paleobot. Ser., homeotype no. 6035.

REFERENCES CITED

Adam, D.P.
  1967   Late Pleistocene and Recent palynology in the central Sierra Nevada, California. *In* E.J. Cushing and H.E. Wright, Jr. (eds.), Quaternary Paleoecology, pp. 275-301. New Haven: Yale Univ. Press.
Axelrod, D.I.
  1941   The concept of ecospecies in Tertiary paleobotany. Nat. Acad. Sci. Proc. 27: 545-551.
  1944a The Mulholland flora. Carnegie Inst. Wash. Pub. 553: 103-146.
  1944b The Oakdale flora. Carnegie Inst. Wash. Pub. 553: 147-166.
  1944c The Sonoma flora. Carnegie Inst. Wash. Pub. 553: 167-206.
  1944d The Pliocene sequence in central California. Carnegie Inst. Wash. Pub. 553: 207-224.
  1950a Further studies of the Mount Eden flora, southern California. Carnegie Inst. Wash. Pub. 590: 73-117.
  1950b The Piru Gorge flora of southern California. Carnegie Inst. Wash. Pub. 590: 158-214.
  1956   Mio-Pliocene floras from west-central Nevada. Univ. Calif. Pub. Geol. Sci. 33: 1-316.
  1957   Age-curve analysis of angiosperm floras. Jour. Paleo., vol. 31: 273-280.
  1958   The Pliocene Verdi flora of western Nevada. Univ. Calif. Pub. Geol. Sci. 34: 91-160.
  1966   The Pleistocene Soboba flora of southern California. Univ. Calif. Pub. Geol. Sci. 60: 1-109.
  1976a History of the conifer forests, California and Nevada. Univ. Calif. Pub. Botany 60: 1-62.
  1976b Evolution of the Santa Lucia fir (*Abies bracteata*) ecosystem. Missouri Bot. Garden. Ann. 63: 24-41.
Axelrod, D.I., and H.P. Bailey
  1968   Paleotemperature analysis of Tertiary floras. Paleogeography, Paleoclimatology, Paleoecology 6: 163-195.

1976 Tertiary vegetation, climate, and altitude of the Rio Grande depression, New Mexico-Colorado. Paleobiology 2: 235-254.

Axelrod, D.I., and W.S. Ting
1960 Late Pliocene floras east of the Sierra Nevada. Univ. Calif. Pub. Geol. Sci. 39: 1-118.

Bachman, S.B.
1978 Pliocene-Pleistocene breakup of the Sierra Nevada-White-Inyo Mountains block and formation of Owens Valley. Geology 6: 461-463.

Bailey, H.P.
1960 A method of determining the warmth and temperateness of climate. Geografisker Annaler 42: 1-16.
1964 Toward a unified concept of the temperate climate. Geogr. Review 54: 516-545.

Bateman, P.C., and C. Wahrhaftig
1966 Geology of the Sierra Nevada. Calif. Div. Mines and Geol. Bull. 190: 107-172.

Christensen, M.N.
1966 Late Cenozoic crustal movements in the Sierra Nevada of California. Geol. Soc. Amer. Bull. 77: 163-182.

Condit, C.
1938 The San Pablo flora of west central California. Carnegie Inst. Wash. Pub. 476: 217-268.
1944 The Table Mountain flora (California). Carnegie Inst. Wash. Pub. 553: 57-90.

Critchfield, W.B.
1971 Profiles of California vegetation. USDA Forest Service Research Paper PSW-76: 1-54.

Curry, R.R.
1966 Glaciation about 3,000,000 years ago in the Sierra Nevada. Science 154: 770-771.
1968 Quaternary climatic and glacial history of the Sierra Nevada, California. Ph.D. thesis, Univ. California, Berkeley. 238 pp.

Curtis, G.H.
1951 The geology of the Topaz Lake quadrangle and the eastern half of the Ebbetts Pass quadrangle. Ph.D. thesis, Univ. California, Berkeley. 310 pp.
1954 Mode of origin of pyroclastic debris in the Mehrten formation of the Sierra Nevada. Univ. Calif. Geol. Sci. 29: 453-502.

Dallimore, W. and A.B. Jackson
1966 A Handbook of Coniferae and Ginkgoaceae, 4th ed. London: Arnold. 729 pp.

Dalrymple, G.B.
1964a Potassium-argon dates and the Cenozoic chronology of the Sierra Nevada, California. Ph.D. thesis, Univ. California, Berkeley. 109 pp.
1964b Cenozoic chronology of the Sierra Nevada, California. Univ. Calif. Pub. Geol. Sci. 47: 1-41.

Durrell, C.D.
1959 Tertiary stratigraphy of the Blairsden quadrangle, Plumas County, California. Univ. Calif. Pub. Geol. Sci. 34: 161-192.
1966 Tertiary and Quaternary geology of the northern Sierra Nevada. Calif. Div. Mines and Geol. Bull. 190: 185-197.

Erwin, H.D.
1934 Geology and mineral resources of northeastern Madera County, California. Calif. Jour. Mines and Geology 30: 7-78.

Evernden, J.F., D.E. Savage, G.H. Curtis, and G.T. James
1964 Potassium-argon dates and the Cenozoic mammal chronology of North America. Amer. Jour. Sci. 262: 145-198.

Gilbert, F.L.
1959 Metamorphism of the Lake Alpine Area, Alpine County, California. M.A. thesis, Univ. California, Berkeley. 70 pp.

Graham, A.
   1976   Late Cenozoic evolution of tropical lowland vegetation in Veracruz, Mexico. Evolution 29: 723-735.
Grant, T. A., J. R. McCleary, and R. L. Blum
   1977   Correlation and dating of geomorphic and bedding surfaces on the east side of the San Joaquin Valley, using dip. In M. J. Singer (ed.), Soil Development, Geomorphology, and Cenozoic History of the Northeastern San Joaquin Valley and Adjacent Areas, California. A Guidebook for the Joint Field Session, Amer. Soc. Agronomy, Soil Sci. Soc. Amer. and Geol. Soc. Amer. pp. 312-318. Univ. California, Davis.
Gray, J., and L. R. Kittleman
   1967   Geochronometry of the Columbia River basalt and associated floras of eastern Washington and western Idaho. Amer. Jour. Sci. 265: 257-291.
Hay, E. A.
   1976   Cenozoic uplifting of the Sierra Nevada in isostatic response to North American and Pacific plate interactions. Geology 4: 763-766.
Huber, N. K., and C. D. Rinehart
   1965a The Devils Postpile National Monument. Calif. Div. Mines and Geol. Mineral Info. Serv. 18 (6): 109-118.
   1965b Geologic map of the Devils Postpile quadrangle, Sierra Nevada, California. U.S. Geol. Surv. Quad. Map GQ-437.
Leighly, J.
   1938   The extremes of the annual temperature march with particular reference to California. Univ. Calif. Pub. Geogr. 6: 191-234.
Lindgren, W.
   1911   The Tertiary gravels of the Sierra Nevada of California. U.S. Geol. Surv. Prof. Paper 73. 226 pp.
Matthes, F.
   1930a Geologic history of the Yosemite Valley. U.S. Geol. Surv. Prof. Paper 160. 137 pp.
   1930b The Devils Postpile and its strange setting. Sierra Club Bull. 15: 1-8.
Noble, D. C., D. B. Slemmons, M. J. Korringa, et al.
   1974   Eureka Valley tuff, east-central California and adjacent Nevada. Geology 2: 139-142.
Piper, A. M., H. S. Gale, H. E. Thomas, and R. W. Robinson
   1939   Geology and ground-water hydrology of the Mokelumne area, California. U.S. Geol. Surv. Water-Supply Paper 780. 230 pp.
Slemmons, D. B.
   1953   Geology of the Sonora Pass region, California. Ph.D. thesis, Univ. California, Berkeley. 201 pp.
   1966   Cenozoic volcanism of the central Sierra Nevada, California. Calif. Div. Mines and Geol. Bull. 190: 199-208.
Sudworth, G. B.
   1908   Forest Trees of the Pacific Slope. U.S. Dept. Agric., Forest Service. 441 pp. (republished by Dover, 1967).
Tanai, T., and J. Wolfe
   1977   Revisions of *Ulmus* and *Zelkova* in the middle and late Tertiary of western North America. U.S. Geol. Surv. Prof. Paper 1026: 1-14.
Vanderhoof, V. L.
   1933   A skull of *Pliohippus tantalus* from the later Tertiary of the Sierran foothills of California. Univ. Calif. Pub. Geol. Sci. Bull. 23: 183-194.
Vasek, F. C.
   1966   The distribution and taxonomy of three western junipers. Brittonia 18: 350-372.

Wang, Chi-Wu
    1961   The Forests of China, with a Survey of Grassland and Desert Vegetation. Maria Moors
            Cabot Foundation Publ. no. 5, Harvard Univ. 313 pp.
Wilshire, H.G.
    1956   The history of Tertiary volcanism near Ebbetts Pass, Alpine County, California. Ph.D.
            thesis, Univ. California, Berkeley. 126 pp.
    1957   Proplytization of Tertiary volcanic rocks near Ebbetts Pass, Alpine County, California.
            Univ. Calif. Pub. Geol. Sci. 32: 243-271.

Chapter II Plates

PLATE 1

Views near the Mt. Reba locality

FIG. 1. View of Mt. Reba ridge from Underwood Valley, looking southwest. Fossil locality is in the saddle on the crest of the ridge.

FIG. 2. Plant-bearing andesitic sandstone interbedded with mudflow breccias in the foreground. The base of the overlying Mt. Reba Conglomerate lies on the slope on the far side of the nearest clump of wind-swept bushes. It extends up the ridge to the sulcus near the snowbank, where it is faulted against the Relief Peak Formation on which the fire lookout is perched. Black, craggy Underwood Mudflow Breccia overlies the granitic basement on the ridge that forms the west side of Underwood Valley. The view continues to the north on Plate 2.

PLATE 2

Geomorphic stages in the Mt. Reba area

View north from the Mt. Reba fossil locality showing the rolling, gentle relief of the Sierran basement on which the Miocene volcanic rocks are perched. The 4,000-foot-deep (1,200 m) gorge of the Mokelumne River in the middle distance represents the Canyon cycle which is cut into the granitic terrain. The broad shallow basin of Underwood Valley below the observer, which is also carved into granodiorite, represents the Mountain Valley stage. Similar "mountain valleys" are visible across the Mokelumne River gorge at the same general level as Underwood Valley.

Both the Mountain Valley and Canyon stages developed after the deposition of the Mt. Reba Conglomerate (5 m.y.?) that unconformably overlies the Disaster Peak Formation (7 m.y.) in which the flora is preserved (see Plate 1).

PLATE 3

Modern Vegetation Related to the Mt. Reba flora

FIG. 1. Douglas fir forest in the valley of Sutter Creek, looking to the north-facing slope. *Pseudotsuga* dominates the middle and lower north-facing slope, and *Quercus chrysolepis* is a regular associate. The equivalent fossil species of both taxa are among the dominants of the flora.

FIG. 2. Douglas fir forest in drainage of Indian Creek, a small tributary of the North Yuba River at an altitude of 2,400 ft. (732 m). The Douglas fir forest is on a northerly-facing slope, and mixed conifer forest reaches down from the ridge-top. *Lithocarpus* and *Quercus chrysolepis* are important subcanopy trees in both forests, and also form sclerophyll vegetation, as seen at right margin of the view.

PLATE 4
Mt. Reba fossils
FIGS. 1-3. *Pseudotsuga sonomensis* Dorf. Hypotype nos. 5954, 5955, 5953.
FIGS. 4-6. *Sequoiadendron chaneyi* Axelrod. Hypotype nos. 5989-5991.

PLATE 5

Mt. Reba fossils

FIG. 1. *Cupressus mokelumnensis* Axelrod. Paratype no. 5969. Cone and branchlets.
FIG. 2. *Typha lesquereuxii* Cockerell. Hypotype no. 5992.
FIG. 3. *Cupressus mokelumnensis* Axelrod. Holotype no. 5967. Cone and branchlets.

PLATE 6

Mt. Reba fossils

FIGS. 1, 3. *Cupressus mokelumnensis* Axelrod. Paratype nos. 5968, 5970.

FIG. 2. *Abies concoloroides* Brown. Hypotype no. 5948.

FIG. 4. *Juniperus* sp. Axelrod. no. 5988.

FIG. 5. *Salix wildcatensis* Axelrod. Hypotype no. 6000.

FIG. 6. *Salix hesperia* (Knowlton) Condit. Hypotype no. 5997.

FIGS. 7, 8. *Salix boisiensis* Smith. Hypotype nos. 5995, 5996.

FIG. 9. *Lithocarpus klamathensis* (MacGinitie) Axelrod. Hypotype no. 6003.

PLATE 7
Mt. Reba fossils

FIGS. 1-5. *Quercus hannibalii* Dorf. Hypotype nos. 6016, 6019, 6020, 6018, 6017.

FIGS. 6-9. *Lithocarpus klamathensis* (MacGinitie) Axelrod. Hypotype nos. 6005, 6002, 6001, 6004.

III

THE TURLOCK LAKE FLORA FROM STANISLAUS COUNTY

Chapter III Contents

PLATES

INTRODUCTION

The Turlock Lake flora comes from localities on two small islands in the western part of Turlock Lake, a reservoir in the lowest Sierran foothills 37 km east of Modesto, in southeastern Stanislaus County (Plate 8, fig. 1) This flora of 25 species adds greatly to our knowledge of the flora and the vegetation in the piedmont belt of the Sierra Nevada during the Late Neogene (Hemphillian). Representing only the second flora to be described from the entire region of the lower Sierran slope, it supplements materially the information supplied by the small Oakdale flora of 15 species, situated 17 mi. (27 km) northwest (Axelrod, 1944c). Only 4 woody species are common to these floras; of the 21 woody species at Turlock Lake, 17 (80 percent) are not now known from the Oakdale flora. As outlined below, the reasons for this are chiefly differences in site, climate, and age. Of additional interest is the contribution that this flora makes to a better under-standing of the ecologic setting under which the rich associated Turlock Lake vertebrate fauna lived. It was during the excavation of this fauna that Dennis C. Garber of Davis, California, collected the leaves on which this report is based. Acknowledgment is due him for his careful collecting of these fragile specimens which he turned over to me for study. They occur at normal reservoir level in wet, thick bentonitic shales which must be carefully excavated and then wrapped in newspaper and slowly dried for several months —otherwise the blocks with the specimens disintegrate to piles of acorn-sized kernels.

PRESENT PHYSICAL SETTING

The fossil flora occurs on islands in Turlock Lake at an elevation of 240 ft. (73 m). The area is dissected into low, broadly rounded hills separated by broad valley flats. Average hill elevation decreases from near 325 ft. (100 m) at the east to about 295 ft. (90 m) along the west margin of the reservoir and falls off progressively farther west. The relief around Turlock Lake has been carved into the Late Cenozoic alluvial deposits of the Pliocene Mehrten and the Pleistocene Turlock Lake and Riverbank formations, all composed of detritus derived from the Sierra Nevada. Natural drainage in the area is controlled by the Tuolumne River, situated a few hundred meters north of Turlock Lake. It flows west to join the San Joaquin River 35 mi. (56 km) distant, which drains north to the estuary at Stockton.

The Turlock Lake area is in the upper part of the Valley Grassland that dominates low elevations in this part of the Central Valley. However, the Tuolumne River floodplain directly north of the reservoir, although now largely under cultivation, supported a woodland-savanna composed chiefly of *Quercus lobata* and scattered clumps of *Populus fremontii*. The steep banks of the river valley, which are entrenched 100-150 ft. (30-45 m) into the terrain, are covered with *Q. wislizenii*, *Q. douglasii*, and various shrubs. As elevation and precipitation rise gradually to the east toward La Grange, grassland gives way to a pure *Q. douglasii* woodland-savanna that covers the hills. Farther east, where precipitation rises above 17.5 in. (450 mm), the woodland is dominated by *Pinus sabiniana, Q. douglasii,* and *Q. wislizenii. Aesculus californica* is commonly scattered in the community, and *Q. lobata* is also prominent, but chiefly in valleys where the water-table is higher. Streambanks in the region support several trees, notably *Acer negundo, Alnus rhombifolia, Faxinus oregona, Populus fremontii,* and *Salix lasiandra.* A local, disjunct stand of *Platanus racemosa* is reported to occur on the Tuolumne River a few miles east of Turlock Lake, near La Grange (Griffin and Critchfield, 1972, map 59). Large shrubs along the river include *Calycanthus occidentalis, Rhamnus californica, Salix exigua, S. lasiolepis,* and others. Scattered on drier sites in the digger pine-oak woodland belt are diverse evergreen shrubs, notably *Adenostema fasciculatum, Arcto-staphylos mariposa, A. viscida, Ceanothus cuneatus, Dendromecon rigida, Heteromeles arbutifolia, Rhamnus californica,* and *R. ilicifolia.* They contribute to patches of chaparral chiefly on slopes with thin soil, but these are basically seral to woodland (Axelrod, 1975).

TABLE 7
Climatic Data for Stations Nearest Turlock Lake

	Modesto		Oakdale Woodward Dam		Denair	
Elevation	27.7	m	65.5	m	37.8	m
Distance from Turlock Lake	37	km	32	km	21	km
Precipitation	307	mm	343	mm	312	mm
Mean annual temperature	15.8°	C	15.9°	C	16.0°	C
Mean July temperature	24.6°		25.4°		25.0°	
Mean January temperature	7.2°		7.1°		7.4°	
Mean annual range	17.4°		18.3°		17.6°	
Warmth (*W*)	14.6°		14.6°		14.6°	
Days warmer than *W*	200.4		200.4		200.4	
Temperateness (or Equability)	54+		53+		54	

The climate of this area is semi-arid, with relatively low rainfall, hot summers, and mild winters. As shown by the data for the nearest meteorological stations (Table 7), mean annual temperature is about 60°F (15.6°C), with mean July temperature 77°F (25°C) and mean January temperature 45°F (7.2°C). This gives a warmth (*W*) of 58°F (14.5°C), or 197 days with a mean temperature rising above 58°F (14.5°C). The temperateness (*M*) or equability of climate is *M* 54, graded on a scale of *M* 100 for an ideal maximum. By comparison, San Francisco, situated 105 mi. (170 km) west-northwest, has a rating of *M* 74. Precipitation at nearby stations averages 12-13 in. (305-330 mm), but it probably is near 14-15 in. (355-380 mm) at Turlock Lake, for it rises to 16 in. (405 mm) at La Grange, situated 6 mi. (10 km) to the northeast at an elevation of 300 ft. (92 m). The climatic data for the stations noted in Table 7 provide a basis for estimating the differences between the present climate and that which is inferred for the fossil flora, as discussed below.

GEOLOGIC OCCURRENCE

The Turlock Lake area is on the east side of the Central Valley of California, in the northern part of its southern sector, the San Joaquin Valley. The Central Valley is a great structural downwarp that lies west of the Sierra Nevada. The downwarp is asymmetrical, sloping gradually to greater depths near the west border of the valley. The relatively complete and continuous section of Cretaceous, Tertiary, and Quaternary strata preserved in the basin is dominantly marine, but continental sediments increase after the Middle Miocene. The maximum thickness of the sedimentary rocks is at the south end of the San Joaquin Valley, where the Tertiary and Quaternary rocks aggregate 28,600 ft. (8,500 m) (Dibblee and Oakeshott, 1953, p. 1502; Hackel, 1966). In that area sedimentation and contemporaneous downwarping have been so rapid that the post-Middle Pliocene continental deposits are over 10,500 ft. (3,500 m) thick (de Laveaga, 1952, p. 102).

The Sierra Nevada was tilted westward in the Late Cretaceous, and its erosional debris was carried into the Central Valley. The older formations on the east side dip more steeply than the younger ones because of the continuous, intermittent tilting of the Sierran block, though the rocks are not folded. It is in the western part of the trough that compression dominates, as shown by the folded and faulted Coast Range terranes. During the Middle and Late Tertiary, the products of volcanism in the northern half of the low Sierran block were transported westward, building up extensive deposits of volcanic and volcaniclastic sediments along the lower flanks of the range and extending into the Central Valley. These deposits were constructed chiefly by rivers which were under the influence of a biseasonal distribution of precipitation, with convectional showers in the warm season and cyclonic disturbances in the colder part of the year. As the finer deposits spread out away from the central river courses, they added much material to the floodplains. Locally in these accumulating finer deposits, conditions were favorable for the preservation of fossil plants and animals, as exemplified by the Turlock Lake flora and fauna.

The geology of the Turlock Lake area has been mapped and discussed by Davis and Hall (1959) and by Mannion (1960). The basement rocks on which the Tertiary sediments rest are exposed 6 mi. (10 km) east, near La Grande. They make up the Jurassic metavolcanics of the Logtown Formation, which is intruded locally by Upper Jurassic granodiorite and related plutonic rocks. The oldest Tertiary formation that lies on the basement is represented by the Ione Formation, composed chiefly of deeply weathered conglomerate, sandstone, and shale. In most areas the finer sediments have been reduced by weathering to anauxite-bearing clays, quartz sand, and a red-brown iron oxide crust, apparently the result of long exposure to weathering in a winter-dry subtropical climate in the Early Eocene. Overlying the varicolored Ione rocks are white rhyolite tuffs and associated sediments of the Valley Springs Formation, which is Lower Miocene as judged from radiometric evidence (21-22 m.y.; Dalrymple, 1964a, 1964b). Above the Valley Springs are the tuffs and volcaniclastic sediments that represent the andesitic Mehrten Formation in which the flora is preserved.

Mehrten Formation

The type area of the Mehrten Formation is 45 mi. (75 km) north, in northeastern San Joaquin County near Clements, at Camanche (formerly Mehrten) Dam (Piper et al., 1939). It forms a distinctive sequence of dark sandstones, claystones, and conglomerates with interbedded tuffs and mudflows, all dominantly of andesitic composition. The debris was carried to the lower flanks of the range from centers of fissure eruption in western Nevada and the Sierra Nevada (Durrell, 1944), thence into the present Central Valley, where it is downwarped, as shown by its occurrence in the subsurface (Piper et al., 1939; Davis and Hall, 1959). More recently, Slemmons (1966) restricted the Mehrten Formation to the section that lies above the Stanislaus Formation, composed of basalt flows and welded tuffs that have since been traced from Nevada across the Sierran summit area into the foothill belt near Knights Ferry (Noble et al., 1974). The andesites and associated volcaniclastic sediments that lie below the Stanislaus Formation represent the Relief Peak Formation, which is about 3,280 ft. (1,000 m) thick in its type area near Sonora Pass (Slemmons, 1966). In areas where the two formations are not separated by the flows of the Stanislaus Formation—which includes most of the Sierran slope—discrimination between them is not easy. In general, the Relief Peak is composed of darker, more basic andesites rich in augite and hypersthene, whereas the Mehrten is made up chiefly of lighter hornblende-rich andesites, as well as the reworked older volcanics.

Davis and Hall (1959) pointed out that the Mehrten Formation can be divided into three units, though their gradational boundaries are at times uncertain. The lowest unit is a scoriaceous to pumiceous sandstone and andesite cobble conglomerate. It is strongly cross-bedded and associated with andesite tuffs and mudflows, fills depressions in the underlying rocks, and varies in thickness from locality to locality. Davis and Hall note that good exposures of this unit are at Knights Ferry, where it reaches a maximum thickness of about 99 ft. (30 m). The contact of the Mehrten (restricted) with the underlying Stanislaus Formation is nicely exposed in cuts on State Highway 120, just west of the Tuolumne-Stanislaus counties boundary. Here the basal hornblende andesite lithic tuff includes large angular blocks of black lavas of the Stanislaus Formation and rests on it. The section grades upward into a fluvio-lacustrine section composed chiefly of alternating andesitic gravels, sandstones, and siltstones that occur in beds from 3-20 ft. (1-6 m) thick, and some persist along the strike for a mile or more. Associated with them are thick tuffaceous sandstones and thin mudflows. This unit, about 395 ft. (120 m) thick, is clearly a floodplain deposit and is represented in the eastern part of the Turlock Lake area. The upper part of the formation forms a poorly exposed sequence of soft clays, silts, and sands, with minor lenses of pebble conglomerate. Its higher parts often include pinkish to flesh-colored clays ranging from 2-4 ft. (0.6-1.2 m) thick. The unit is well exposed along the Stanislaus River east of Oakdale, and on the shores of the western part of Turlock Lake as well. The Mehrten increases in thickness to 1,200 ft. (365 m) in the subsurface west of Modesto, as indicated by the deep-drilling data presented by Davis and Hall (1959, p. 10, fig. 3).

The ages of the andesitic formations in this area are now well established. Following the stratigraphy proposed by Slemmons (1966), the age of the upper part of the Relief Formation in the foothill belt that underlies the Stanislaus Formation (K/Ar age = 9.5 m.y.; Noble et al., 1974) is indicated by the Two-Mile Bar fauna. Its mammals have been correlated with those in the upper Mint Canyon, Neroly, Orinda, Columbia, Coal Valley, and other faunas (Stirton and Goeriz, 1942), several of which have now been dated radiometrically in the range of 10-11 m.y. (Evernden et al., 1964). On the other hand, the Mehrten Formation (restricted) that overlies the Stanislaus, or the Relief Peak, where it is absent, is largely Middle Pliocene (Hemphillian), as judged from the Oakdale fauna recovered from the upper Mehrten Formation at Schell Ranch (Stirton and Goeriz, 1942) and Tulloch Ditch (Vanderhoof, 1933). This fauna is similar in age to those in the Pinole Tuff (dated at 5 m.y.), Etchegoin, and Petaluma formations, all considered Middle and Late Hemphillian by Stirton and Goeriz (1942), and the older part of which ranges down to about 9 m.y. (Evernden et al., 1964). A Middle Pliocene (Hemphillian) age for the Oakdale flora (Axelrod, 1944c), preserved just below the middle of the Mehrten Formation southwest of Knights Ferry, has been suggested on the basis of its composition and climatic indications. The age of the lowest part of the Mehrten Formation (restricted) is not closely established, but probably is mid-Hemphillian or slightly older. The rich mammalian fauna from the plant-bearing beds at Turlock Lake, which occurs in the upper 115-130 ft. (35-40 m) of the Mehrten Formation, is Late Hemphillian, as judged from the relations of its taxa to those in other Hemphillian faunas, notably the Pinole (dated at 5 m.y.), as outlined by Wagner (1976). As reviewed below, paleoclimatic evidence suggests that the Turlock Lake fauna and flora may be somewhat younger, or about 4.5 to 4.0 m.y.

In the Mokelumne River drainage to the north of Turlock Lake, the Mehrten Formation is overlain unconformably by the Laguna Formation. It is judged to be Late Pliocene on the basis of the record of a tooth of *Neohipparion* cf. *gidleyi* recovered from the Bottomore Well, and considered to be from the Laguna Formation according to Piper et al. (1939, p. 60). The Laguna Formation is composed chiefly of light-colored sands, silts, and a few pebble beds that were largely derived from a granitic terrane. The Laguna Formation has not been recognized in the Tuolumne River area, where the Mehrten is overlain unconformably by the Turlock Lake Formation along a contact that lies just west of the reservoir (Davis and Hall, 1959). The Turlock Lake Formation was also derived chiefly from a granitic terrane, and a major unconformity separates it from the Mehrten. Turlock Lake is generally conceded to be of Pleistocene age, and according to Arkley (1962) contains fine rock flour that implies glaciation in the Sierran uplands.

In summary, the Turlock Lake flora and associated vertebrate fauna occur in the uppermost part of the Mehrten Formation and are thus younger than the Oakdale flora from the middle of the Mehrten, in the nearby region to the north. Stratigraphic relations and the ages of mammalian faunas in the nearby area indicate that the Turlock Lake biota are no older than Lake Hemphillian (5 m.y.). As outlined below, paleoclimatic evidence suggests they may be somewhat younger, perhaps 4.0 to 4.5 m.y.

COMPOSITION OF THE TURLOCK LAKE FLORA

As now known, the Turlock Lake flora is made up of 25 species, distributed among 1 conifer, 4 monocots, and 20 dicots. Two of the monocots, *Cyperacites* and *Juncus*, are referred only to genus, because they are not represented by systematically diagnostic specimens. Two new species of *Ceanothus* are represented in the flora, and 3 taxa previously referred to other species are given new names. All the others have been described earlier.

Systematic List of Species

Pinaceae
  Pinus sturgisii Cockerell
Cyperaceae
  Cyperus sp.
Juncaceae
  Juncus sp.
Liliaceae
  Smilax remingtonii new species
Typhaceae
  Typha lesquereuxii Cockerell
Salicaceae
  Populus garberii new species
  Salix edenensis Axelrod
  Salix hesperia (Knowlton) Condit
  Salix laevigatoides Axelrod
Fagaceae
  Quercus dispersa (Lesquereux)
  Axelrod
  Quercus douglasoides Axelrod
  Quercus pliopalmerii Axelrod
  Quercus wislizenoides Axelrod

Lauraceae
  Persea coalingensis (Dorf) Axelrod
  Umbellularia salicifolia (Lesquereux)
  Axelrod
Platanaceae
  Platanus paucidentata Dorf
Rosaceae
  Prunus turlockensis new species
Fabaceae
  Amorpha condonii Chaney
Anacardiaceae
  Toxicodendron (*Rhus*) *franciscana*
  Axelrod
Rhamnaceae
  Ceanothus tuolumnensis new species
  Ceanothus turlockensis new species
  Rhamnus moragensis Axelrod
  Rhamnus precalifornica Axelrod
Ericaceae
  Arbutus matthesii Chaney
Oleaceae
  Forestiera buchananensis Condit

As listed in Table 8, the fossil plants are very similar to living ones. The typical habit or life-form of the taxa, as judged from their closest living allies, shows that the flora is composed of 8 trees, 13 shrubs, 3 herbaceous perennials, and 1 or 2 vines. The 10 evergreen dicots in the flora, marked by an asterisk (*) in Table 8, include 4 trees and 6 shrubs. It is especially noteworthy that only two seeds of pine have been encountered in the collection. This probably reflects the fact that most conifers lived in the hills farther east, where precipitation was higher and temperature lower than at Turlock Lake. Nonetheless, remains of a fossil digger pine allied to *Pinus sabiniana* might be expected to have had an appreciable record in the flora, for it regularly occurs today with many of the taxa that have close allies in the Turlock Lake assemblage.

As judged from the representation of specimens (Table 9), the most abundant remains are those of *Quercus wislizenoides* and *Populus garberii*, and 2 others, *Platanus paucidentata* and *Salix hesperia*, are relatively common. This is expectable, for 3 of these 4 genera (*Platanus, Populus, Salix*) are confined largely to the borders of streams and lakes. The occurrence in the flora of 3 other taxa, *Cyperus, Juncus*, and *Typha*, whose nearest allies are found almost exclusively along the margins of streams, lakes, and ponds, is also significant in this respect, for they provide additional evidence regarding the physical setting in which the flora lived. The abundance of the oak is explainable in part by its very hard, durable leaves, which would favor their easy preservation, and in part by its water requirements. *Q. wislizenii*, which is similar to the dominant oak, is confined to river valleys in areas where precipitation is below 15-17.5 in. (380-450 mm), as can be observed today on the Tuolumne River a few hundred yards north of the fossil

TABLE 8

Living Species Most Similar to the Turlock Lake Taxa
(arranged according to their usual life-form or habitat)

Fossil species	Similar living species
Trees	
*Arbutus matthesii	A. menziesii
*Persea coalingensis	P. podadenia
Pinus sturgisii	P. ponderosa
Platanus paucidentata	P. racemosa
Populus garberii	P. tremula
Quercus douglasoides	Q. douglasii
*Quercus wislizenoides	Q. wislizenii
*Umbellularia salicifolia	U. californica
Shrubs	
Amorpha condonii	A. californica
*Ceanothus tuolumnensis	C. sorediatus
*Ceanothus turlockensis	C. palmerii
Forestiera buchananensis	F. neomexicana
Prunus turlockensis	P. Asiatic aff.
*Quercus dispersa	Q. dumosa
*Quercus pliopalmerii	Q. dunnii
*Rhamnus moragensis	R. ilicifolia
*Rhamnus precalifornica	R. californica
Salix edenensis	S. exigua aff.
Salix hesperia	S. lasiandra
Salix laevigatoides	S. laevigata
† Toxicodendron franciscana	T. diversiloba
Perennial Herbs	
Cyperus sp.	C. spp.
Juncus sp.	J. spp.
Typha lesquereuxii	T. angustifolia
Vine	
Smilax remingtonii	S. californica
† Toxicodendron franciscana	T. diversiloba

*Evergreens.   †May be either a vine or a shrub.

sites. Only in areas where precipitation exceeds 17.5-19 in. (450-500 mm) does it also occur regularly in the oak woodland-savanna away from river courses.

Two other taxa in the flora that are moderately frequent, *Umbellularia* with 6 specimens and *Amorpha* with 4, also prefer moist sites. The remaining 18 species are represented by 3 specimens or less, and 11 of them are known from only a single specimen.

PALEOECOLOGY

The Turlock Lake flora can best be interpreted in the light of its depositional environment and the ecologic relations of living species most similar to the fossil plants.

TABLE 9

Representation of Specimens in the Turlock Lake Flora

Species	Loc. T-14	Loc. T-20	Totals
Quercus wislizenoides	58	128	186
Populus garberii	8	137	145
Platanus paucidentata	15	13	28
Salix hesperia	6	14	20
Umbellularia salicifolia	. . .	6	6
Amorpha condonii	1	3	4
Salix laevigatoides	1	2	3
Quercus pliopalmerii	. . .	3	3
Pinus sturgisii	. . .	2	2
Quercus douglasoides	2	. . .	2
Persea coalingensis	1	1	2
Cyperus sp.	1	. . .	1
Juncus sp.	. . .	1	1
Typha lesquereuxii	1	. . .	1
Smilax remingtonii	1	. . .	1
Salix edenensis	1	. . .	1
Quercus dispersa	. . .	1	1
Prunus turlockensis	. . .	1	1
Toxicodendron franciscana	1	. . .	1
Rhamnus precalifornica	. . .	1	1
Rhamnus moragensis	. . .	1	1
Ceanothus tuolumnensis	. . .	1	1
Ceanothus turlockensis	. . .	1	1
Arbutus matthesii	1	. . .	1
Forestiera buchananensis	. . .	1	1
Total Specimens	98	317	415
Total Species	14	18	25

Depositional Setting

The upper Mehrten Formation in the western part of the Turlock Lake area is chiefly lacustrine, with thin and generally fine fluvatile interbeds. The bulk of the sediments were deposited in a quiet lake, as indicated by their very fine grain, rhythmic bedding, and the graded bedding. Also, remains of the Sacramento blackfish (*Orthodon*) have been recovered from the plant beds at Locality T-14 (Casteel and Hutchinson, 1973). They not only indicate a lowland area, but one typified by shallow lakes and sluggish sloughs. Furthermore, *Orthodon* is reported to avoid running water like that in most streams (Murphy, 1950). The preponderantly fine sediments and their distribution over a wide area indicate a setting in a region of very low relief. It was essentially a featureless plain, far more subdued than the area is at present.

The fossil sites are now at an elevation of 230 ft. (70 m) above sea level, but were certainly somewhat lower during deposition, probably near 115 ft. (35 m.) The setting at a very low elevation, on a broad and nearly featureless plain that had relatively large lakes, is consistent with the occurrence of the remains of *Smilodonichthys* in the deposit (Cavender and Miller, 1972). This giant salmonoid is estimated to have been 190 cm (about 6 ft.) long. It is unique in that it has numerous (over 100) gill rakers, and its jaws and palate do not have feeding teeth, all features compatible with feeding in quiet water, and a pelagic habit for the most part (Cavender and Miller, 1972).

The fossil plants evidently accumulated at some distance from the shore. This is suggested by the plant-bearing sediments, some of which appear to have been transported as turbidity flows into the lake. Not only are the sediments often graded, many of the leaves lie at diverse angles to the bedding planes, as though their positions were determined by complex movements in the transporting agent. Hence it is expectable that they represent a mixture of taxa that inhabited more than one environment along the shores of the lake and entering river, probably the ancestral Tuolumne.

The question arises as to the location of the nearest hills. The flora is now situated 12 km west of the metavolcanics of the Logtown Formation that form the Sierran basement in this area. During deposition, the area between the fossil site and the basement terrane was a broad floodplain with essentially no relief, and the Mehrten sediments reached eastward into the present area of the metamorphic terrane, as is evident from their projected dip close to the basement in the SE 1/4 sec. 12, T35, R13E. From the area of the basement outcrops eastward, relief now rises rapidly to the east, attaining a summit level of 2,950 ft. (900 m) on Penon Blanco 19 mi. (30 km) east of Turlock Lake. This ridge is one of a number of parallel ridges of metamorphic rocks along the lower Sierran foothill belt that most investigators believe have formed a discontinuous belt of abrupt hills in the foothill area since the Early Cretaceous (e.g., Turner, 1894; Turner and Ransome, 1897; 1898; Lindgren, 1911, pp. 37-39, 197-199, and 218-219; Wahrhaftig, 1962). However, considerable evidence suggests that they were not as high during the Tertiary as the present hills, which are largely exhumed and have been rejuvenated by faulting and possibly by warping. In the first place, to judge from currently accepted rates of erosion, they could not have persisted for fully 110 million years at nearly their present levels: they would have been reduced to a plain in 15 m.y. (Gilluly, Waters, and Woodford, 1968, p. 78). Second, the Mehrten and Valley Springs formations extend as piedmont deposits for fully 25 miles (40 km) southeast of Turlock Lake, along the front of the Sierran basement (see Rogers, 1966). Examination of sedimentary structures in the Mehrten Formation in the area from Turlock Lake southeast toward Merced Falls indicates that its sediments were derived from the east and northeast, from the general drainage area of the present central Tuolumne and southern Stanislaus rivers. This implies transportation *across* the present line of steep metamorphic hills at the front of the range. Third, in the region of the type area of the Mehrten Formation, the metavolcanic rocks that make up the Bear Mountain-Hogback Mountain ridge have been exhumed. These hills were largely surrounded by Mehrten sediments which have since been stripped over a wide area, as may be seen in the hills between New Hogan Reservoir and Pardee Reservoir (Valley Springs and San Andreas quadrangles). By inference, they have also been stripped in the area to the south, in the region of the drainages of the central Tuolumne and Stanislaus rivers. At the end of deposition, the Mehrten sediments stood at least 500-600 ft. (150-180 m) higher, giving the basement hills in the Bear Mountain-Hogback Mountain area a maximum relief of 1,000-1,500 ft. (300-450 m) on the Mehrten plain, as compared with over 2,000 ft. (600 m) today. Fourth, some of the present relief is certainly the result of renewed movement along the foothill fault system following Mehrten deposition, as indicated by detailed mapping of the gravels (Goldman, 1964). Marchand (1977) also notes that the foothill fault system was reactivated during the Cenozoic, as shown by substantial offsets in the Stanislaus Formation (9 m.y.) near Sonora and by one or more faults north and east of Merced which appear to have offset the Ione Formation and also broken the Late Pliocene China Hat Member of the Laguna Formation into rotated blocks, thus altering the regional west dip. This agrees with evidence that the China Hat gravel, composed of a great flood of metamorphic debris, is interbedded in the upper part of the Laguna Formation and thus implies rapid uplift in the Pliocene (Late Blancan). Further, Marchand reports that Pliocene? gravel deposits near Oroville that seem correlative with the China Hat Member also show important offset. In addition, Marchand (1977), as well as Guacci and Purcell (1978) and Page et al. (1978), review evidence that the numerous northwesterly-trending fractures in the eastern San Joaquin and Sacramento valleys coincide with patterns in the basement rocks, and that some of them in the Late Pliocene Laguna and Early Pleistocene Turlock Lake and Riverbank formations show displacements of 3-6 ft. (1-2 m), and that displacements are also in Holocene soils. The data are incomplete, yet the evidence clearly suggests that the Logtown and allied metamorphic formations probably formed only low ridges in the lower foothill belt when the Mehrten sediments were deposited, that the volcaniclastics largely buried them, and that they were later exhumed and uplifted by faulting, and perhaps also by warping. Principal uplift of the basement highs that mark the foothill belt probably occurred chiefly during the Late Pliocene (the Laguna Formation dips much less than the Mehrten) and Early Pleistocene; it has continued to the present, and therefore was essentially contemporaneous with the Coast Range orogeny.

The position of the Turlock Lake area with respect to the estuary system that occupied the San Joaquin Valley has not been established. As charted on the Geologic Map of California, Santa Cruz Sheet (Jennings and Strong, 1959), and also by Reed (1933), Woodring, Stewart, and Richards (1940), and Hackel (1966), the Etchegoin Sea that occupied much of the southern San Joaquin Valley reached northward along the east margin of the Diablo uplift to near latitude 36° 30'. The marine deposits are replaced to the east and north by nonmarine strata derived from sediment whose source was chiefly Sierran. That estuarine conditions may have extended northward toward the Turlock Lake area may be inferred from the ecologic requirements of the fish in the upper Mehrten Formation, for they inhabit quiet water; at least they indicate lakes not far from sea level, and these may have drained to nearby estuaries. Whether estuarine conditions reached up the central San Joaquin Valley toward the Turlock Lake area may be determinable also from subsurface data. However, this task was not undertaken because of the difficulty of obtaining the original well-log data, but particularly because of the uncertainty of interpreting the depositional environments they may represent.

In summary, the Turlock Lake flora lived on the shores of a lake situated on the broad floodplain of upper Mehrten deposition. The area was one of very low relief and close to sea level. Drainage probably was south into the inland sea that occupied the Coalinga region at this time. Evidence suggests that the high metavolcanic basement ridges that now dominate the area east of the flora had not yet been elevated, and that the region was a low plain rising gradually eastward, blanketed by a thin veneer of Mehrten Formation.

Vegetation

As judged from the ecologic occurrences of the nearest living relatives of the fossil species, two principal plant communities are represented, oak woodland-savanna and floodplain vegetation. As discussed below, there are also indications that broadleaved sclerophyll forest lived near at hand. Some of the taxa that are typical of the oak woodland belt that covered the interfluves no doubt occurred also with the floodplain species on the principal drainageways. These are indicated by an asterisk (*).

OAK WOODLAND-SAVANNA
  Trees: *Arbutus matthesii, Quercus douglasoides,* **Q. wislizenoides,* **Umbellularia salicifolia*
  Shrubs/small trees: *Ceanothus tuolumnensis, C. turlockensis, Quercus dispersa,* **Rhamnus moragensis, R. precalifornica,* **Toxicodendron franciscana*
FLOODPLAIN
  Trees: *Persea coalingensis, Platanus paucidentata, Populus garberii,* **Quercus wislizenoides,* **Umbellularia salicifolia*
  Shrubs/small trees: *Amorpha condonii, Forestiera buchananensis, Prunus turlockensis,* **Rhamnus precalifornica, Salix edenensis, S. hesperia, S. laevigatoides, Toxicodendron franciscana*
  Vines: *Smilax remingtonii,* †*Toxicodendron franciscana*
  Herbaceous perennials: *Cyperus* sp., *Juncus* sp., *Typha lesquereuxii*

†As judged from the nature of its close living relative, *Toxicodendron diversiloba, T. franciscana* was both a shrub and a vine.

Most of the species in the flora that make up these communities are represented by closely allied plants in nearby parts of the Sierra Nevada at slightly higher elevations where precipitation is greater. The flora along the Stanislaus River at levels near 490-650 ft. (100-200 m) in the vicinity of Knights Ferry and eastward, situated 16 mi. (25 km) north of Turlock Lake, includes the following species that have close equivalents (*) or ecologic counterparts (†) in the fossil flora.

Aesculus californica	**Quercus wislizenii*
Alnus rhombifolia	**Rhamnus californica*
†*Ceanothus cuneatus*	**Rhamnus ilicifolia*
Fraxinus oregona	**Toxicodendron diversiloba*
Heteromeles arbutifolia	*Symphoricarpos albus*
†*Populus fremontii*	*Salix exigua
**Quercus douglasii*	*Salix laevigata
*Quercus dumosa	*Salix lasiandra
Quercus lobata	*Vitis californica*

Umbellularia is not listed here because it is not in the lower Sierran foothill belt. It occurs farther east, generally near 1,475-2,000 ft. (450-610 m) throughout the central foothills of the Sierra (see Griffin and Critchfield, 1972, map 84). Also, *Arbutus* does not inhabit the lower part of the range, but occurs at medium levels, 2,500-3,000 ft. (760-920

m) and generally farther north, in the lower mixed conifer forest and the upper part of the woodland belt where precipitation exceeds 31 in. (800 mm). Another plant with an equivalent in the fossil flora that is not in the lower foothill belt is *Ceanothus palmerii*. It commonly frequents the middle mountain slopes in southern California and occurs also in the central Sierra Nevada in Amador and Eldorado counties at the lower margin of forest and in the upper woodland belt. In this regard, it is recalled that there are 2 seeds in the flora that appear to represent yellow pine, a species which regularly straggles down into the upper woodland belt where it is associated with the taxa above. The record of *Smilax* provides another indication of a climate moister than that now in the lower woodland belt, or at Turlock Lake, for it occurs in the Coast Ranges from Napa County northward, reaching into the Klamath-Siskiyou region and thence into south-western Oregon. *Smilax* also occurs along the Sacramento River near Tehama, a site to which its seeds may have been carried from the mountains to the west. In any event, it occurs there with *Platanus racemosa*, another important plant in the present-day flora that is not in the central Sierran foothills. *Platanus* lives at low elevations along the Sacramento River in the central part of the state, rising to somewhat higher levels in the woodland belt of the southern Sierra Nevada. It is in the Coast Ranges, where winters are more mild, that *Platanus* regularly reaches well up into the upper part of the broadleaved sclerophyll belt—above the lower oak woodland-savanna zone—to meet mixed conifer forest.

Available evidence suggests that environment in the Coast Ranges is more nearly like that represented by the fossil flora than is that of the woodland belt in the Sierra Nevada. For example, along Atascadero Creek southwest of Atascadero, vegetation also shows relationship to the fossil flora (Plate 8, Fig. 2). The woodland belt that dominates the lowlands is composed of *Quercus agrifolia, Q. douglasii, Q. lobata,* and *Pinus sabiniana. Q. agrifolia* largely replaces *Q. wislizenii* in this region, which has a milder climate than the Sierran foothill area. Oak woodland ranges westward to meet the broadleaved sclerophyll forest, the taxa of which reach down the valley and occur on cooler slopes within a few feet of the oak woodland. Among the taxa in this area that are closely associated are:

Acer macrophyllum	†*Quercus agrifolia*
Alnus rhombifolia	*Quercus douglasii*
Amorpha californica	*Quercus dumosa*
Arbutus matthesii	*Quercus lobata*
Ceanothus cuneatus	*Rhamnus californica*
†*Ceanothus leucodermis*	*Rhamnus ilicifolia*
Ceanothus sorediatus	*Salix exigua*
†*Cornus glabrata*	*Salix laevigata*
Platanus racemosa	*Salix lasiandra*
†*Populus fremontii*	*Toxicodendron diversiloba*
Populus trichocarpa	*Umbellularia californica*
Prunus ilicifolia	

*Close living representative.
†Ecologic equivalent of fossil species.

Apart from these 17 species that have closely similar taxa in the fossil flora, it is note-worthy that *Forestiera neomexicana* occurs in the nearby area, along the Salinas River.

There it is a common associate of *Platanus racemosa, Populus fremontii, Quercus douglasii, Q. lobata,* and several species of *Salix.* Furthermore, *Quercus dunnii* (= *Q. palmerii*) is also in this region, for a disjunct population from southern California occurs in the woodland belt in Peachy Canyon a few miles west of Paso Robles. This makes a total of 18 of the 21 woody plants in the flora that have similar or ecologically equivalent species in this rather limited area.

Similar vegetation is also present along the Nacimiento River west of Jolon, situated 20 mi. (32 km) southwest of King City, where the following have been observed (see Plate 15, Fig. 2, in Chapter IV).

Adenostema fasciculata	**Quercus douglasii*
Aesculus californica	*Quercus dumosa
Alnus rhombifolia	*Quercus lobata*
**Amorpha californica*	**Quercus wislizenii*
**Arbutus menziesii*	**Rhamnus californica*
Arctostaphylos glauca	**Rhamnus ilicifolia*
†*Ceanothus integerrimus*	*Ribes menziesii*
**Ceanothus leucodermis*	*Ribes sanguineum*
Cercocarpus montanus	*Salix exigua
°*Pinus ponderosa*	*Salix laevigata
**Platanus racemosa*	*Salix lasiandra
†*Populus fremontii*	*Salix lasiolepis*
Populus trichocarpa	**Toxicodendron diversiloba*
Prunus ilicifolia	**Umbellularia californica*
Quercus agrifolia	

*Represented by a very similar species in the Turlock Lake flora, or
†by an ecologic equivalent of *P. garberii* or *C. palmerii.*
°Occurs in sclerophyll belt at Ponderosa Camp, on the river to the west.

The view shown in Plate 15, Figure 2 (see Chapter IV) was taken at the site of the Nacimiento weather station west of Hunter-Liggett airfield. Rainfall here is 29 in. (735 mm) annually. The occurrence in this restricted area of some 17 woody taxa that have allies or ecologic equivalents in the fossil flora provides a general measure of the nature of the conditions under which the assemblage lived.

The relations of vegetation in the outer Coast Ranges, notably in the Santa Lucia Range as well as to the north and south, differ considerably from those in the Sierra Nevada. A broadleaved sclerophyll forest, which occupies a belt fully 3,200 ft. (1,000 m) in elevation in the Santa Lucia Range, is absent in the Sierra Nevada. The zone is reduced there to only occasional isolated pockets of sclerophyll forest adjacent to conifer forest, to which its species also contribute. The chief taxon is *Quercus chrysolepis,* which occurs in the Coast Ranges in the middle and upper part of the sclerophyll forest belt. The nature of the Turlock Lake flora suggests that a richer broadleaved sclerophyll forest occupied the lower slopes of the Sierra into the later Pliocene. It is recalled that the Late Miocene Table Mountain flora (Condit, 1944b) has a number of species that occur today chiefly in the broadleaved sclerophyll belt, notably:

Fossil Species	Similar Living Species
Arbutus matthesii Chaney	*A. menziesii*
Cercis buchananensis Condit	*C. occidentalis*
Cercocarpus antiquus Lesquereux	*C. betuloides*
Forestiera buchananensis Condit	*F. neomexicana*
Pinus pretuberculata Axelrod	*P. attenuata*
Platanus paucidentata Dorf	*P. racemosa*
Quercus dispersa (Lesq.) Axelrod	*Q. dumosa*
Umbellularia salicifolia (Lesq.) Axelrod	*U. californica*

In addition, the Remington Hill flora, somewhat younger in age, also shows that broadleaved sclerophyll vegetation was present in the region (Condit, 1944a). Among the present-day taxa that live in this belt, the following occur in the Remington Hill flora:

Fossil Species	Similar Living Species
Arbutus matthesii Chaney	*A. menziesii*
Arctostaphylos martzii Condit	*A. manzanita*
Forestiera buchananensis Condit (recorded as *Ceanothus precuneatus*)	*F. neomexicana*
Lithocarpus klamathensis (MacG.) Axelrod (recorded as *Quercus simulata* Kn.)	*L. densiflorus*
Platanus paucidentata Dorf	*P. racemosa*
Populus prefremontii Dorf	*P. fremontii*
Prunus petrosperma Condit	*P. ilicifolia*
Quercus douglasoides Axelrod	*Q. douglasii*
Quercus prelobata Axelrod	*Q. lobata*
Quercus wislizenoides Axelrod	*Q. wislizenii*
Salix hesperia (Kn.) Condit	*S. lasiandra*

This sclerophyll vegetation lived on warmer slopes bordering a *Sequoia-Chamaecyparis* forest and its associated mesic taxa that included species of *Aesculus, Crataegus, Liquidambar, Magnolia, Persea, Platanus,* and *Ulmus,* as well as the sclerophyllous trees that also occupied the moister valley. Inasmuch as the broadleaved sclerophyll forest formed a major vegetation zone in western Nevada into the late Miocene (Axelrod, 1956), its presence in the Sierra Nevada is not surprising. That it no longer occurs there as a major vegetation zone results chiefly from the present climate, which is notably less equable than that in the Coast Ranges. Equally important, the sclerophyll forest that was in the area during the later Pliocene probably succumbed to the heavy snows during the successive glacials, for it is now clear that they severely damage the evergreens (Axelrod, 1976).

There are only 3 taxa in the fossil flora that are now represented by closely similar species in California today. *Prunus turlockensis* appears to be related to Eurasian species as judged from the intricate, peach-like sculpturing of the seed coat. *Populus garberii* is closely allied to *P. tremula* of Eurasia, and more distantly to *P. grandidentata* of eastern North America. In terms of the flora, it has an ecologic counterpart in *P. fremontii*, which inhabits the margins of streams throughout the California woodland belt. *Persea* is the other taxon that is not now native to California. It probably was a regular member of the floodplain vegetation, and also reached up into the woodland belt in moister areas at a time when summer rainfall was still present in the region.

Climate

From the preceding observations, it is evident that the Turlock Lake flora indicates that rainfall was higher in the lowest foothill area of the Sierra Nevada than it is at present. An estimate of not less than 25 in. (635 mm) rainfall, as compared with about 13 in. (330 mm) today, seems indicated. It probably was not much higher; otherwise taxa from the upper parts of the oak woodland belt and the adjacent forest, notably *Acer macrophyllum, Arctostaphylos mariposa, Ceanothus integerrimus, Lithocarpus densiflorus, Pseudotsuga menziesii,* and *Quercus chrysolepis,* might be expected to have contributed some of their structures to the accumulating record. That there was some summer rainfall is implied by *Populus garberii,* for it is allied to poplars now in eastern Asia and the eastern United States. In addition, *Persea coalingensis* also represents an alliance whose relatives are now restricted to areas with summer rainfall. Both species are frequent in the Pliocene of the Kettleman Hills situated 120 mi. (195 km) south, where they are in the Etchegoin and San Joaquin formations. It is also recalled that taxa with summer-rain requirements are in the nearby Oakdale flora, where *Populus* (cf. *mexicana*), *Robina* (cf. *neomexicana*), and *Sapindus* (cf. *drummondii*) are recorded (Axelrod, 1944c), and also in the Mulholland to the west (Axelrod, 1944b), where there are species of *Acer, Karwinskia, Nyssa, Populus, Quercus,* and *Sapindus,* whose nearest relatives are now in summer-rain areas only. Both floras are somewhat older, and therefore have a larger number of taxa that are no longer represented in the region.

Comparisons of the thermal requirements of modern vegetation and taxa most nearly allied to those in the Turlock Lake flora suggest that temperature was milder than that now along the lower slopes of the Sierra (Table 10). Summer temperatures evidently were more nearly like those now in the Coast Ranges. Winters were mild, probably with only rare light frost, to judge from the presence of frost-sensitive *Persea* in this flora and from its abundance in the nearby Coalinga region to the south, at the same time. A thermal regime not greatly unlike that near the inner limits of avocado groves in southern California today (Fig. 10), and with some summer rain, may be inferred for the flora. This implies that mean temperature may have been slightly higher than that now in the southern Coast Ranges, and that the range of temperature was much less than that presently in the Turlock Lake area. The available evidence suggests a climate with the following general characteristics:

Annual precipitation		Temperature	F	C
Total amount	25 in. 635 mm	Mean annual	60°	15.5°
Seasonal occurrence	winter & summer	Mean July	72.5°	22.0°
		Mean January	48.5°	9.0°
		Warmth (*W*) or Effective		
		Temperature (*ET*)	58.5°	14.7°
		Days warmer than *W* (*ET*)	204	204
		Temperateness (or equability)	*M* 60	*M* 60

TABLE 10

Climatic Data for Stations that Provide a Means
for Estimating the Conditions Under Which
the Turlock Lake Flora Lived
(see Figure 10 for the thermal parameters)

Station	Elevation (m)	Precip. (mm)	Mean temperature (°C)			Range (°C)
			July	Jan.	Annual	
Sierra Nevada						
1. Camp Pardee	200	521	26.0	7.6	16.5	18.4
2. Electra Power House	221	762	25.5	7.7	16.2	17.8
3. Folsom Dam	107	622	25.6	7.2	16.2	18.4
4. Mokelumne Hill	427	782	24.9	6.8	14.8	18.1
5. Sonora	557	813	25.4	6.6	15.4	18.8
6. Rocklin	73	587	25.6	6.8	15.7	18.8
7. Oroville 7 SE	162	645	26.1	7.1	16.2	19.0
South Coast Ranges						
8. Atascadero	255	445	19.8	6.5	13.0	13.3
9. Dry Creek Reservoir	460	343	23.8	9.4	16.3	14.4
10. Gilroy	320	721	21.3	7.9	133.6	13.4
11. Paso Robles	251	363	22.1	7.8	14.7	14.3
12. Nacimiento Dam	235	518	22.6	7.2	15.4	15.4
13. Pinnacles Nat. Mon.	399	630	23.0	7.9	15.2	15.1
14. San Antonio Mission	322	—	—	—	15.7	15.8
15. Santa Margarita	304	747	22.7	6.1	14.7	16.6
Coastal Southern California						
16. Bonita	32	305	21.0	11.5	16.2	9.5
17. Corona	216	315	23.7	10.9	17.1	12.7
18. Escondido	201	412	23.0	10.6	16.6	12.4
19. La Mesa	161	328	22.3	11.9	17.0	10.4
20. Ojai	229	513	22.9	10.1	16.3	12.8
21. Santa Paula	80	345	20.5	11.9	16.3	8.6
22. Tustin	36	358	21.9	11.2	16.3	10.7
23. Santa Barbara	31	448	19.7	11.4	15.7	8.3

Comparison with the data in Table 7, which presents the general climate in the
Turlock Lake area today, provides a basis for estimating the degree of climatic change in
the region. Precipitation has been reduced by about half, from 25 in. to 13 in. (635 mm to
335 mm); mean annual temperature has increased only slightly, from 59.9°F to 60.8°F
(15.5°C to ~ 16°C); the annual range of temperature has increased, from 55.9°F to
63.5°F (13.3°C to 17.5°C); warmth of climate has scarcely changed (*W* 58.4°F, or
14.7°C, or 204 days with mean temperature warmer than 58.4°F, or 14.7°C); but
equability has decreased measurably (*M* 60 to *M* 54). These changes reflect the shift to a
more continental climate, as the Coast Ranges to the west were elevated to give the 300

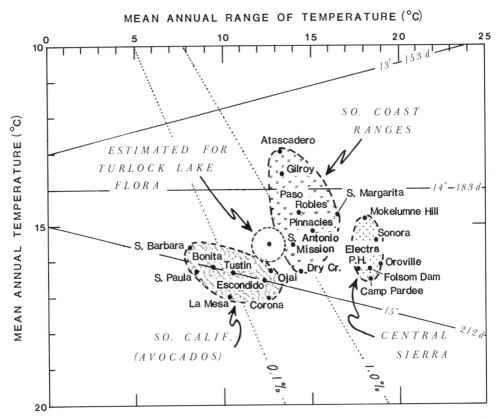

FIG. 10. Inferred thermal field for the Turlock Lake flora is circled. The low temperature for Atascadero reflects its position opposite the mouth of a major canyon, down which cold air drains from the higher Coast Ranges to the west.

mile-long (480 km) Central Valley its interior position. This brought to it a warmer and drier climate and greater ranges of temperature than those present prior to the Coast Range disturbance. This event occurred chiefly during the Early and Middle Pleistocene, as shown by rocks of these ages which have been deformed significantly, the evidence for which is summarized by Christensen (1966).

COMPARISON WITH THE OAKDALE FLORA

The Oakdale flora is situated just below the middle of the Mehrten, whereas Turlock Lake is close to the top of the formation. The time that separates them may be on the order of 0.5 to 1 m.y., to judge from the nature of the sediments and the differences in floral composition. Age is partly responsible for the floristic differences, though site was no doubt important in giving Turlock Lake a more mesic aspect. The Oakdale flora is associated with thick tuffs, conglomerates, and coarse tuffaceous sandstones laid down on a floodplain that was coursed by turbulent waters. Between times of flood, it was well-drained and relatively dry. That these volcaniclastic sediments provided dry sites for the Oakdale flora is consistent with the occurrence there of *Heteromeles, Mahonia, Quercus douglasoides,* and *Ribes mehrtensis,* and the general absence there of the more mesic plants that are at Turlock Lake—notably its species of *Arbutus, Ceanothus, Persea, Pinus, Smilax,* and *Toxicodendron*—or quite rare at Oakdale but common at Turlock Lake—notably *Umbellularia.*

The Turlock Lake flora accumulated in a lake and its taxa largely inhabited the shore area. Although coarse pebbly sandstone and cobble conglomerate are below the flora on the eastern shore 2-3 mi. (3-5 km) distant, most of the section is fine-grained. A high water-table probably accounts in part for its generally mesic aspect, with its dominant poplar, sycamore, and willows. However, its generally mesic species of *Arbutus, Ceanothus* (cf. *palmerii, sorediatus*), *Persea, Pinus, Smilax, Toxicodendron,* and *Umbellularia* seem to reflect more fundamental differences, notably a higher annual precipitation.

The contrasts between these nearby Pliocene floras, which are not very different in age, can best be appreciated when they are compared as follows:

Turlock Lake	Oakdale
Pinaceae	
Pinus sturgisii	
Cyperaceae	
Cyperus sp.	
Juncaceae	
Juncus sp.	
Smilaxaceae	
Smilax remingtonii	
Salicaceae	Salicaceae
Populus garberii	*P. alexanderii (trichocarpa)*
Salix edenensis	*P. pliotremuloides (tremuloides)*
Salix hesperia	*P. parce-dentata (mexicana)*
Salix laevigatoides	*S. wildcatensis (lasiolepis)*
Fagaceae	Fagaceae
Quercus dispersa	*Q. dispersa (dumosa)*
Quercus douglasoides	*Q. douglasoides (douglasii)*
Quercus pliopalmerii	
Quercus wislizenoides	*Q. wislizenoides (wislizenii)*
	Ulmaceae
	Celtis kansana (reticulata)
	Berberidaceae
	Mahonia cf. *marginata (japonica)*
Lauraceae	Lauraceae
Umbellularia salicifolia	*Umbellularia salicifolia (californica)*
Persea coalingensis	
Platanaceae	
Platanus paucidentata	

Saxifragaceae
  Ribes mehrtensis (quercetorum)
Rosaceae   Rosaceae
  Prunus turlockensis   *Heteromeles sonomensis (arbutifolia)*
Fabaceae   Fabaceae
  Amorpha condonii   *Robinia californica (neomexicana)*
Anacardiaceae
  Toxicodendron franciscana

                                      Sapindaceae
                                        Sapindus oklahomensis (drummondii)
Rhamnaceae   Rhamnaceae
  Ceanothus mokelumnensis

                                        (*Ceanothus precuneatus*
                                        (unidentifiable scarp, rejected)

  Ceanothus turlockensis
  Rhamnus precalifornica
  Rhamnus moragensis
Ericaceae   Ericaceae
  Arbutus matthesii   *Arctostaphylos oakdalensis (mariposa)*
Oleaceae
  Forestiera buchananensis

It is apparent that only 7 of the 18 families in the two floras are in common, and that the taxa which make up the Fagaceae are nearly alike. The other families in common are for the most part represented by different species or genera. Thus, the Salicaceae have a total of 8 species, but none common to both floras. In the Lauraceae, *Umbellularia* is shared by both floras, but *Persea* is not at Oakdale. In the Fabaceae, the genera are different— *Robinia* at Oakdale and *Amorpha* at Turlock Lake—and the Ericaceae are also represented by different genera—*Arbutus* at Turlock Lake and *Arctostaphylos* at Oakdale. In addition, the Rosaceae are represented by different genera—*Prunus* at Turlock Lake, *Heteromeles* at Oakdale.

The general ecologic requirements of the taxa in the Turlock Lake flora seem for the most part to indicate a climate that was definitely moister than that at Oakdale. This is consistent with its younger age and with the general trend of precipitation in this region. As outlined earlier (Axelrod, 1944b, 1944d, 1944e, 1948, 1957a, 1957b, and 1971) there was a significant rise in rainfall in the Pliocene. That this trend was regional in scope has also been suggested, and the new evidence provided by the floras in the foothill belt lends further support to this idea.

THE LATE NEOGENE RAINFALL TREND

A relatively complete sequence of Late Neogene floras from the San Francisco Bay region has provided a basis for reconstructing the rainfall trend in that area (Axelrod, 1944e, 1957a, 1957b, 1971). Although some of the floras have a more coastal position (i.e., Santa Rosa) than others (i.e., Mulholland), there is nonetheless clear evidence that there was a gradual decrease in precipitation into the late Miocene (Middle Hemphillian), followed by a sharp rise in Pliocene (Late Hemphillian-Early Blancan) time.[2] The evidence is provided by a sequence of floras which shows that the dominant Late Miocene lowland live-oak sclerophyll vegetation was replaced in the Pliocene by taxa representing a coast redwood forest, a change that implies a significantly wetter climate. As suggested earlier (Axelrod, 1944e, p. 215), the marked rise in precipitation heralds the buildup of the first major ice sheets in high latitudes. Current evidence suggests that in the Northern Hemisphere the last lengthy period of stable sea level and low ice volume terminated about 3.2 m.y. ago. Since then the earth has been subject to the stress of continually fluctuating climate, with conditions like those of the present occurring infrequently (Shackleton and Opdyke, 1977). It is also clear that the rise in precipitation during the Pliocene corresponds to the triggering of the older glaciations in the Sierra Nevada. These have now been dated by their stratigraphic position with respect to volcanic rocks. The Deadman Till (Curry, 1966, 1968) lies between volcanic rocks dated at 3.0 and 2.7 m.y. The McGee till (Blackwelder, 1931; Putnam, 1962) in the nearby region, which has a similar discordant topographic position and displays a comparable high degree of weathering, rests on basalt dated at 2.74 m.y. There is no doubt that they are essentially of the same age, but whether they represent the same glaciation probably cannot be established, owing to the nature of the rock samples available for dating.

The data that can be used to reconstruct a rainfall trend in the Sierra Nevada are not as complete as those in the coastal area to the west. Nonetheless, there is no evidence at present to suggest that the trend in the Sierra Nevada differed from that in the coastal sector (Fig. 11). The oldest Late Neogene flora now known from the lower Sierran slope is the Table Mountain (Condit, 1944b). It is not younger than 10 m.y. (Slemmons, 1966; Dalrymple, 1963, 1964; Noble et al., 1974). The flora occurs in sediments that have a stratigraphic position similar to the *Nannippus* cf. *tejonensis* Merriam recovered from Springfield Shaft in the nearby area (in Condit, 1944b), and which occurs in the Neroly, Coal Valley, and other faunas that are dated at ~ 10 m.y. (in Evernden et al., 1964). The rainfall estimate of 25-30 in. (635-760 mm) by Condit (1944b) is revised upward here, because a number of the taxa which have equivalents in the flora—in the genera *Carya*, *Gledtischia*, *Ilex*, *Magnolia*, *Persea*, and *Platanus*—reach their western limits of distribution in the central United States in areas where precipitation usually is well above 30 in.

The Remington Hill flora from the northern part of the Sierra was earlier considered contemporaneous with the Table Mountain flora (Condit, 1944a). The evidence that it is younger rests on its more modern composition as compared with the Denton Creek flora situated 40 mi. (65 km) north-northeast in the Penman Formation, the basal part of which is dated at 10.8 m.y. (Dalrymple, 1964a). The Remington Hill is considerably younger than a large flora from the Bonta Formation near Webber Lake, situated 25 mi. (40 km) northeast of Remington Hill. The more mesic aspect of the Remington Hill, as compared with the Table Mountain flora, is evident from its *Sequoia-Chamaecyparis* forest with deciduous hardwoods (*Acer, Liquidambar, Ulmus*) which occupied a valley

---

2. It is recalled that recent changes in the placement of the Miocene-Pliocene boundary (from 10 or 12 m.y. up to 5 m.y.) account for the differences in age recorded in this report as compared with earlier ones.

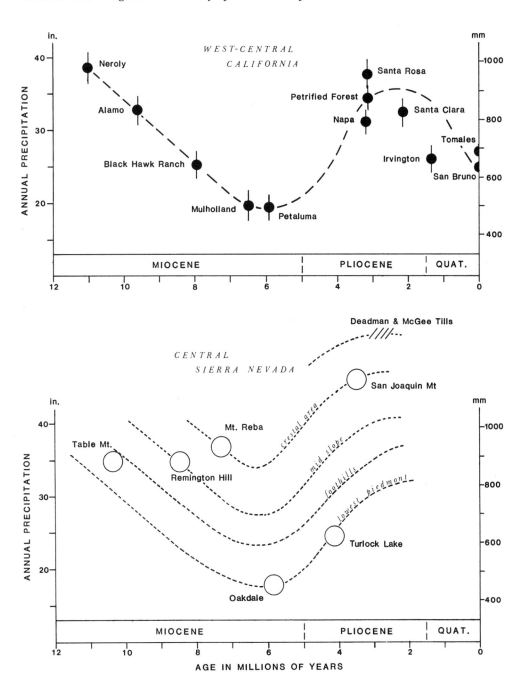

FIG. 11. Estimates of rainfall indicated by some Neogene floras in the Sierra Nevada, compared with the rainfall trend constructed earlier from the more complete sequence in the San Francisco Bay region (Axelrod, 1971, fig. 2).

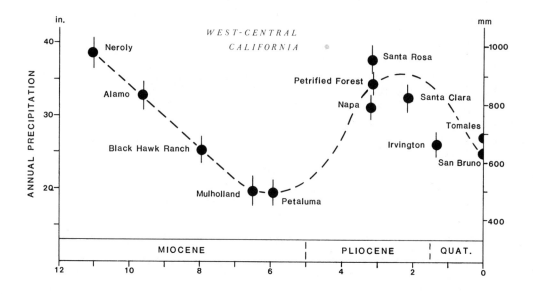

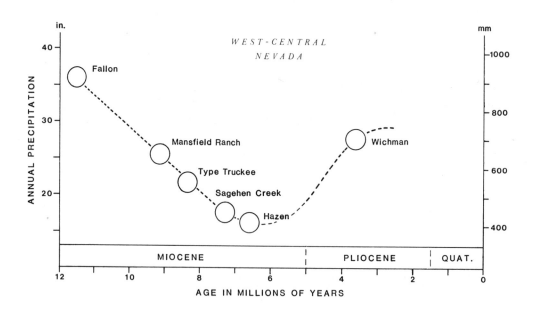

FIG. 12. Estimates of rainfall indicated by some Neogene floras in west-central Nevada and the San Francisco Bay region (compare with Fig. 11).

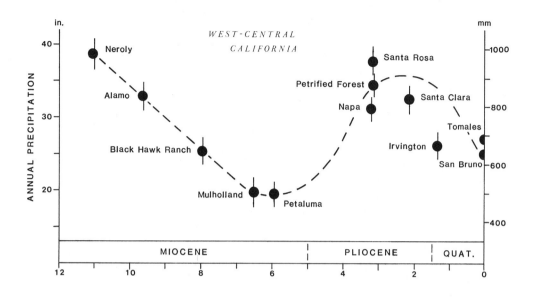

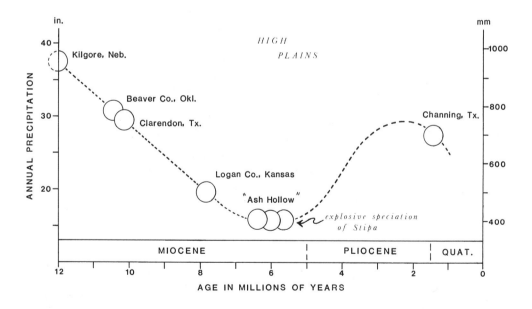

FIG. 13. Estimates of precipitation indicated by some Neogene floras in the High Plains compared with that in the San Francisco Bay region (compare Figs. 11 and 12).

adjacent to broadleaved sclerophyll vegetation of *Arbutus, Lithocarpus*, and several species of *Quercus*. The difference, as compared with the Table Mountain flora, reflects its occurrence higher up on the windward slopes of the Sierra and 95 mi. (150 km) farther north, where climate was more humid—as it is today. Whereas the Table Mountain flora is estimated to have had an elevation of 500 ft. (150 m) the Remington Hill was closer to 1,500 ft. (450 m), according to Condit (1944a, 1944b).

In view of these differences in environmental setting, and hence in the floras and the vegetation that they represent, estimates of precipitation trends in the Sierra must be charted on separate curves, as shown in Figure 11. Clearly, if comparisons are made between floras at different latitudes, or elevations, or position with respect to the east and west sides of an active volcanic range, as has been done by some investigators, the results must lead inevitably to a series of spurious climatic fluctuations. The nature of the problem is nicely exemplified in the present case (Fig. 11). If a precipitation curve was drawn for all these floras without regard to their elevation and geographic position, the erroneous conclusion could be drawn that there was essentially no change in precipitation in this area during the Late Miocene (Table Mountain, Remington Hill, Mt. Reba floras), after which precipitation decreased catastrophically (Oakdale, 6 m.y.), and then rose sharply in the Pliocene (4.5-4 m.y.).

It is the younger Oakdale and Turlock Lake floras that provide critical information with respect to the climatic trend on the low piedmont of the Sierra Nevada. The Oakdale flora is known to be older than the Turlock Lake, since it is at a lower stratigraphic level in the Mehrten Formation. The Turlock Lake and Oakdale mammalian faunas, both of which occur high in the upper part of the Mehrten Formation, have been correlated with the Pinole fauna, dated at 5 m.y. That these upper Mehrten faunas may be slightly younger, say 4.5-4.0 m.y., is wholly probable. If so, then the age of the Turlock Lake fauna can well be 4.0 m.y., because its precipitation then falls on the curve near 25 in. (635 mm), which agrees with the paleoecologic evidence reviewed above. The Turlock Lake is certainly more mesic in aspect than the Oakdale flora, and this is probably not to be attributed solely to difference in site—lake vs. floodplain. Not only does the Turlock Lake flora have more mesic species (*Arbutus, Ceanothus, Persea, Pinus, Smilax, Umbellularia*), the first 5 are not known at Oakdale, and the latter is represented there by only a scrap. Furthermore, the Oakdale has a gooseberry (*Ribes mehrtensis*) with very small leaves which is allied to the semi-arid *R. quercetorum* that lives chiefly in the drier parts of the woodland belt in the southern Sierra Nevada and south Coast Ranges, and is disjunct to central Arizona and Baja California. *Populus* (cf. *mexicana*), *Robinia*, and *Sapindus* are additional Oakdale taxa whose nearest relatives are in Arizona and adjacent Mexico, and also suggest a warmer climate.

The Turlock Lake flora, with an estimated precipitation of not less than 25 in. (635 mm), thus appears to be about 4.0-4.5 m.y. as judged from the rise in precipitation recorded by the floral sequence in the San Francisco Bay region. That sequence was approaching its peak near 3.5 m.y. This is the age of the floras from the Sonoma Volcanics, which are Early Blancan, as compared with the Late Hemphillian Turlock Lake flora. When a flora of Early Blancan age is discovered in the Sierran foothill area, it may be expected to show a further rise, accompanied by a decrease in summer temperature, as implied by the early glaciations in the high Sierra Nevada (Fig. 11).[3]

_____

3. The San Joaquin flora is not included in this discussion because it is in a different floristic province, on

The general rainfall trend constructed for west-central California and the partial trend inferred for the Sierra Nevada are similar to those sketched earlier for a series of small floras in the Carson Sink region of western Nevada and on the High Plains (Axelrod, 1957a, 1957b), and which are revised here using new data (Figs. 12 and 13). Note that the Kilgore flora of western Nebraska (MacGinitie, 1962), even though north of those in Kansas-Oklahoma-north Texas, readily fits into the regional pattern because it is in the same floristic province. As for drought, apart from the retreat of forest, and the trend to smaller leaves shown by the Kansas-Oklahoma floras from the Clarendonian into Hemphillian time (Chaney and Elias, 1936), Elias (1942) documented an outburst of evolution of numerous new species of *Stipa* adapted to semi-arid climate on the High Plains in Hemphillian (Ash Hollow) time.

That the trend to dry climate reached its nadir in the Middle Hemphillian can scarcely be doubted, for the floras are well dated, and a similar trend is indicated in widely separated areas by floras of very different composition. That the trend to drought culminated 5-6 m.y. ago naturally raises the question of what relation this trend—which was regional in scope—may have had on the Mediterranean "salinity crisis" (Hsu et al., 1977), which is dated at 6.2-5.2 m.y. (Kaneps, 1979; Smith, 1977), the time of driest climate. Although salt accumulation during the Messinian (or at any other time) depended first upon a closed basin, an increasingly drier climate would have aided the accumulation of a thick salt section. Analysis of the numerous fossil floras that border the Mediterranean basin should provide critical evidence regarding the trend of land climate and illuminate further the climate of the Messinian, which was contemporaneous with the spread of the Antarctic ice sheet (Shackelton and Kennett, 1975).

The rise in precipitation following the "aridity crisis" also requires explanation. That it may have resulted from closing of the Panamanian portal seems likely. With more warm water now shunted northward on the western sides of the Atlantic (Kaneps, 1979) and Pacific, a slight rise in sea-surface temperature in the middle and higher latitudes would occur, and is in fact reflected in the planktonic (Kennett et al., 1979) and shallow-water molluscan faunas (Durham, 1950, fig. 3). A warmer sea-surface would also bring increased precipitation in summer, as well as more moderate winters. That this occurred may be inferred from the spread of previously relict taxa that required summer rain (*Persea, Pterocarya, Ulmus*), some of which (*Persea*) are frost-sensitive, as documented by the Sonoma floras from western California (Axelrod, 1944d, 1950). In the more mountainous areas at higher latitudes, snow would now accumulate and ice would build up and thence spread.

SUMMARY

The Turlock Lake flora of 25 species is preserved in the uppermost part of the Mehrten Formation, associated with a vertebrate fauna of Late Hemphillian (4-5 m.y.) age. The flora inhabited the shores of a lake and its entering streams and is therefore dominated by taxa that required a high water-table, notably *Platanus, Populus, Salix,* and *Umbellularia*. Abundant oaks and associated sclerophyllous shrubs imply that an

---

the shores of a sea where more equable climate favored the persistence of numerous relict taxa. Apart from the Middle Miocene Temblor flora (15 m.y.), the younger floras in this area are too small to permit close estimates of their rainfall requirements.

oak woodland-savanna covered the drier interfluves on the lowest Sierran piedmont, where grassland dominates today. Representatives of broadleaved sclerophyll vegetation (*Arbutus, Ceanothus, Quercus, Umbellularia*) probably lived in nearby cooler sites, and suggest conditions like those now in the Coast Ranges, where oak woodland and broadleaved sclerophyll vegetation meet today. Only 3 taxa no longer have close equivalents in California today, and both suggest some summer rainfall, a feature indicated by other floras of comparable age in the nearby region.

Past climate in the Turlock Lake area differed from present conditions in that total precipitation has been halved, summer rainfall has been eliminated, and the range of temperature has increased, but mean temperature has scarcely changed. The shift to a more continental climate occurred during the Quaternary, as the Coast Ranges were elevated to the west. As compared with slightly older floras in the region, the Turlock Lake flora records a significant rise in precipitation. This reversal of the Late Neogene trend to decreased precipitation, which was regional in scope, probably was triggered by the closing of the Panamanian portal and was a prelude to the buildup of glaciers in Late Hemphillian time.

SYSTEMATIC DESCRIPTIONS

Family PINACEAE
Pinus sturgisii Cockerell
(Plate 9, figs. 1, 2)

Pinus sturgisii Cockerell, Amer. Jour. Sci. Ser. 4, vol. 26, p. 538, text fig. 2, 1908.

Pinus florissantii MacGinitie, Carnegie Inst. Wash. Pub. 599, p. 84, pl. 18, fig. 12; pl. 20, figs. 1, 3, 4, 1953 (not pl. 19, fig. 2, which remains *P. florissantii*).

Pinus leaves MacGinitie, *ibid.* 599, pl. 18, fig. 1, 1953.

Pinus florissantii Lesquereux. Axelrod, Univ. Calif. Pub. Geol. Sci. vol. 33, p. 276, pl. 4, figs. 19, 20; pl. 17, figs. 10, 11, 1956; Axelrod, *ibid.* 34, pl. 17, fig. 3, 1958; Axelrod, *ibid.*, 39, p. 227 (not pl. 42, fig. 9, which is *P. balfouroides* Axelrod; only specimen no. 8008, which is a fascicle), 1962.

Pinus ponderosa Wolfe (not Lawson), U.S. Geol. Surv. Prof. Paper 454-N, p. N15; pl. 1, figs. 1, 4; pl. 8, figs. 32, 33, 1964.

The type specimen of *P. florissantii* Lesquereux, from the transitional Eo-Oligocene (35 m.y.) Florissant flora of Colorado has long been considered to be a yellow pine allied to the living *P. ponderosa* Lawson (MacGinitie, 1953). It is amply clear, however, that the scales of the cone are not those of a member of subsect. Ponderosae, but of sect. Strobus (Little and Critchfield, 1969). The cone is readily matched by those of the living *P. flexilis* James, which is a regular member of the mixed conifer forests from Colorado to Nevada and eastern California, notably in the Sweetwater Mountains north of Bridgeport and in the San Bernardino Mountains of southern California, and locally along the east front of the Sierra Nevada. Cones produced by the species in the mixed conifer forests are larger than those from subalpine and timberline sites, and are remarkably like the fossil *P. florissantii*. This disposition of *P. florissantii* Lesquereux makes it necessary to apply another name to a number of fossils previously referred to it, and which represent a pine allied to *P. ponderosa*. The name *Pinus sturgisii* Cockerell has been chosen because it is the oldest available epithet.

Cockerell compared the fascicle of *P. sturgisii* with those produced by the living *P. taeda* Linnaeus of the southeastern states. There is a general relation, yet *P. ponderosa* has needles as long and as slender as those of *taeda*. Further, the winged seeds in the Florissant flora are larger than those usually produced by *taeda* and are quite similar to the winged seeds of *P. ponderosa*, as noted earlier by MacGinitie. Thus it seems likely that the Florissant fossils represent a pine more nearly allied to *ponderosa* than to *taeda*. On this basis, the name *sturgisii* is herewith applied to fossils that are allied to the living *ponderosa*.

As for the species listed in the synonymy above, the cone in the Chalk Hills flora previously identified as *P. florissantii* Lesquereux is not a yellow pine either. The cone scales are not pyramidal, and do not have prominent umbos or bristles. A latex impression of the cone, when compared with modern cones of subsect. Balfourianae, which includes *P. aristata*, *P. balfouriana*, and *P. longavea*, clearly shows that it is most nearly like *balfouriana*. This is especially apparent in the structure of the cone scales, which give the appearance of drooping breasts. Furthermore, the umbos are neither prominent nor pyramidal in shape, and they are not terminated by sharp bristles, as in *aristata* and *longaeva*. To emphasize the close relationship between the fossil and the living foxtail pine, the name *balfouroides* is herewith chosen.

The Chalk Hills locality near Virginia City, Nevada, which yielded this fossil, is nearly midway between the present populations of *P. balfouriana*. They are discontinuous between the Marble Mountains-Yollo Bolly Mountain region of northwestern California and the Mt. Whitney region of the southern Sierra Nevada, about 300 miles (483 km) distant. Furthermore, *Abies shastensis*, which is also discontinuous between these areas, is represented by a closely allied fossil in the nearby Purple Mountain flora in the Truckee River gorge (Axelrod, 1975).

The record of pine in the Turlock Lake flora is based on two seeds that are oblong-ovate, 7.0 to 7.5 mm long and 3.5 mm broad. They are rounded at each end, slightly keeled on one side, and the surface is moderately ridged longitudinally. The exocarp is quite thin. The wing is missing, having been lost during transport. These seeds compare quite favorably with those of *P. ponderosa*. Acknowledgment is due Norah Van Kleeck, Director of the Seed Laboratory, Food and Agriculture Department, State of California, for recognizing their affinity. The general absence of needles or cones of yellow pine in the Turlock Lake flora is explicable by their area-weight ratio, which militates against transport for any great distance in a lake or quiet water. The seeds probably were rolled by a current along the bottom of the lake for some distance from the tree (or trees) that produced them.

Collection: U.C. Mus. Pal., Paleobot. Ser., Turlock Lake, hypotype nos. 6036, 6037.

Family CYPERACEAE
Cyperus sp.
(Plate 9, fig. 6)

A fragment of a leaf referred to *Cyperus* is 1.2 cm long and 5 mm wide. It is deeply keeled and has numerous very fine parallel veins, but otherwise shows no distinctive features that would warrant its description as a species. Although it is referred to *Cyperus*, it also shows relationship to leaves of *Scirpus*. The species of these genera live chiefly along the shores of lakes and marshes or slow streams, an environment compatible with the setting under which the flora appears to have accumulated.

Collection: U.C. Mus. Pal., Paleobot. Ser., Turlock Lake, no. 6038.

Family JUNCACEAE
Juncus sp.
(Plate 9, fig. 4)

A slender leaf blade 5.2 cm long and 1.5 mm wide is referred to this genus because it is hollow and round, not angular and solid in cross-section as in a needle of pine. It is like the leaves of various species of *Juncus* which inhabit wet areas today. The specimen is not sufficiently unique to warrant describing it as a species. Its ecologic significance is that species of *Juncus* inhabit wet areas today, and hence support the general environment which has been reconstructed for the area from other lines of evidence.

Collection: U.C. Mus. Pal., Paleobot. Ser., Turlock Lake, no. 6039.

Family LILIACEAE
Smilax remingtonii new name
(Plate 9, fig. 5)

Smilax diforma Condit, Carnegie Inst. Wash. Pub. 553, p. 40, pl. 8, fig. 3, 1944.

The specimen figured by Condit from the Remington Hill flora is much smaller than those of *S. diforma* from the Neroly flora to which he referred it. The leaf is well within the range of normal variation of the living *S. californica* (A. DC.) Gray, and is a closer match for it than is *S. rotundifolia* Linnaeus of the eastern United States with which Condit compared it.

The basal half of a cordate leaf in the present collection is also similar to those produced by the living *S. californica*, and is therefore referred to the Remington Hill species, which is hence renamed *S. remingtonii*. *S. californica* is a frequent vine on stream borders in the broadleaved sclerophyll (oak-madrone) vegetation in the northern Coast Ranges, distributed from Napa County northward through the Klamath-Siskiyou region into southern Oregon, and occurring also locally on the Sacramento River in Butte and Tehama counties.

Collection: U.C. Mus. Pal., Paleobot. Ser., Turlock Lake, hypotype no. 6040.

Family TYPHACEAE
Typha lesquereuxii Cockerell
(Plate 9, fig. 3)

Typha lesquereuxi Cockerell, Amer. Mus. Nat. Hist. Bull. 24, p. 79, pl. 10, fig. 46, 1908; MacGinitie, Carnegie Inst. Wash. Pub. 599, p. 91, pl. 69, fig. 2, 1953.

The distal part of a leaf of *Typha* measuring 5.7 cm long, 6 mm wide below, and tapering to 4 mm above is referred to this species. There are 10 parallel veins below, decreasing upward to 6 in the distal part of the specimen. There is no midrib. The specimen closely resembles the distal end of a cattail leaf, and appears to be more nearly like *T. angustifolia* Linnaeus than *T. latifolia* Linnaeus. *Typha* inhabits quiet water on the margins of lakes, streams, and ponds, an environment similar to that reconstructed for the Turlock Lake flora.

Collection: U.C. Mus. Pal., Paleobot. Ser., Turlock Lake, hypotype no. 6041.

Family SALICACEAE
Populus garberii new name
(Plate 10, figs. 1-3)

Populus balsamoides Goeppert. Lesquereux, U.S. Geol. Surv. Terr., vol. 8, p. 248, pl. 55, fig. 4 only (not figs. 3 and 5, which are *P. emersonii* Condit), 1883; Condit, Carnegie Inst. Wash. Pub. 476, p. 254, pl. 4, fig. 3; pl. 5, figs. 1, 3, 1938.
Populus washoensis Brown. Axelrod, *ibid.* 553, p. 98, pl. 22, figs. 1, 2, 1944.

A new name is needed for the poplars referred earlier by Condit (1938) to the European *P. balsamoides* Goeppert because of their spatial separation and the doubtfulness of their identity. Those in the Neroly material are in no way allied to the living *P. balsamifera*; they differ in shape, in fundamental venation, and in the nature of the marginal teeth.

The fossils are abundantly represented in the Neroly flora, as noted by Condit. In addition, they make up one of the dominant species of the Turlock Lake flora, and are also common in the Black Hawk Ranch flora (Axelrod, 1944a), as well as in the Middle Pliocene Etchegoin Formation of the Kettleman Hills (U.C. Mus. Pal.). All of these fossils represent a poplar, as judged from the relatively medium angle of the basal primaries, and the numerous large teeth. The fossils in the Neroly flora were compared by Condit (1938) with the leaves of the modern *P. grandidentata* of the eastern United States. There is certainly a relation, but the Eurasian *P. tremula* is a closer match in terms of marginal teeth. Like the fossils, the teeth are numerous and not as large nor as widely spaced in the fossil as in *grandidentata*. The species is not allied to the *fremontii-deltoides* plexus, for the species of that group have primaries that have a high angle of divergence and are deltoid, not ovate, in shape.

The leaves figured by Chaney (1938, pl. 6, fig. 4, and pl. 7, figs. 1c, 1d) as *P. pliotremuloides* Axelrod are very similar to those of *P. garberii*, but differ from them in their consistently smaller size and larger teeth. They seem more nearly matched by leaves of the variety *davidiana* than *P. tremula* proper, and have been transferred to *P. subwashoensis* Axelrod (Axelrod, 1956). The abundant leaves in the Black Hawk Ranch flora identified earlier by Axelrod (1944a) as *P. washoensis* Brown are herewith transferred to *P. garberii* on the basis of the more numerous teeth than those normally present in *P. washoensis*, as illustrated by fossils in the 49 Camp (LaMotte, 1936), Blue Mountains (Chaney and Axelrod, 1959), and Chalk Hills (Axelrod, 1962) floras.

It is a pleasure to name this species for Dennis C. Garber, who collected the entire Turlock Lake flora during his excavation for fossil vertebrates which occur in the same beds.

Collection: U.C. Mus. Pal., Paleobot. Ser., Turlock Lake, hypotype nos. 6042-6044, homeotype nos. 6045-6060.

Salix edenensis Axelrod
(Plate 9, fig. 7)

Salix edenensis Axelrod, Carnegie Inst. Wash. Pub. 590, p. 101, 1950 (see synonymy).

A single, small linear leaf in the Turlock Lake flora is assigned to this species on the basis of its close similarity to leaves of the modern sandbar willows, notably such species as *S. exigua*, *S. hindsiana*, and *S. interior*, which range widely in California and throughout the far West.

Collection: U.C. Mus. Pal., Paleobot. Ser., Turlock Lake, hypotype no. 6069.

Salix hesperia (Knowlton) Condit
(Plate 9, figs. 8 and 9)

Salix hesperia (Knowlton) Condit, Carnegie Inst. Wash. Pub. 553, p. 41, pl. 4, fig. 7, 1944 (see synonymy and discussion); Axelrod, *ibid.*, p. 132, 1944 (see synonymy and discussion).

The large leaves of a willow allied to the living *S. lasiandra* are common in the Turlock Lake flora. Most of them are fragmentary, because the tuffaceous clays readily break up, and complete specimens, which would measure 12 to 15 cm, cannot be recovered. The fossils have numerous secondaries that diverge at high angles and loop up gently along the margin, which is either entire or finely toothed, features displayed by the living *S. lasiandra*.

This small tree is widely distributed throughout California in areas outside the desert and below the upland fir forests. It ranges northward to Alaska and eastward into the northern Rocky Mountains.

Collection: U.C. Mus. Pal., Paleobot. Ser., Turlock Lake, hypotype nos. 6061-6062; homeotype nos. 6063-6068.

Salix laevigatoides Axelrod
(Plate 10, fig. 4)

Salix laevigatoides Axelrod, Carnegie Inst. Wash. Pub. 590, p. 55, pl. 2, fig. 10, 1950.

Three specimens in the flora are referred to this species, which is characterized by long slender leaves, numerous camptodrome secondaries, serrate margin, acute bases, and moderately tapering tips. They are smaller than those of *S. lasiandra*, they tend to be wider below rather than near the middle, and the tips are not so acuminate. Among living species, leaves of the red willow, *S. laevigata*, are very similar to the fossil material. This is a widely distributed species at middle to lower elevations along stream banks from the woodland and semidesert shrub communities up into the borders of the mixed conifer forest.

Collection: U.C. Mus. Pal., Paleobot. Ser., Turlock Lake, hypotype no. 6070, homeotype nos. 6071-6072.

Family FAGACEAE
Quercus dispersa (Lesquereux) Axelrod
(Plate 11, fig. 11)

Quercus dispersa (Lesquereux) Axelrod, Carnegie Inst. Wash. Pub. 516, p. 97, pl. 7, figs. 8-9, 1939; Axelrod, *ibid.* 590, p. 147, pl. 2, fig. 7, 1950.

A single specimen and its counterpart seem referable to this species, which falls well within the wide range of variation shown by leaves of the living *Q. dumosa*, the common scrub oak of California. The leaf is generally oblong, with shallow lobes and a rounded base and venation that are readily matched by the leaves of the modern taxon. *Q. dumosa* is a frequent member of the chaparral and occurs also in the oak woodland of the south Coast Ranges and on the islands off southern California.

Collection: U.C. Mus. Pal., Paleobot. Ser., Turlock Lake, hypotype no. 6073; homeo-type no. 6074.

Quercus douglasoides Axelrod
(Plate 11, fig. 8)

Quercus douglasoides Axelrod, Carnegie Inst. Wash. Pub. 553, p. 198, pl. 37, figs. 7-10, 1944. Condit, *ibid.*,
  p. 42, pl. 5, figs. 6, 7, 1944; Axelrod, *ibid.* 590, p. 205, pl. 5, fig. 4, 1950.

The specimens representing this species include medium-sized leaves with irregularly shallow lobes, and the lobes have sharp spiny teeth. Others are nearly oblong and smaller, and have a thick heavy petiole. All of them can readily be matched by leaves of the living blue oak, *Q. douglasii* Hooker and Arnott. It is a regular member of the digger pine-oak woodland that encircles the Central Valley of California, and occurs also in the inner Coast Ranges. It prefers the drier sites, as compared with its associates *Q. wislizenii* and *Q. lobata*, and may form pure stands over the lower foothills well below *Pinus sabiniana* and the other oaks, in a much drier and hotter region.

Collection: U.C. Mus. Pal., Paleobot. Ser., Turlock Lake, hypotype no. 6075; homeo-type nos. 6076-6077.

Quercus pliopalmerii Axelrod
(Plate 12, figs. 5-8)

Quercus pliopalmerii Axelrod, Carnegie Inst. Wash. Pub. 476, p. 174, pl. 5, figs. 1-3, 1936; Axelrod, *ibid.* 516,
  p. 99, 1939.

The record of this species is based on a few slightly asymmetrical ovate leaves and their counterparts. All display a number of distinctive features that ally them with the living *Q. dunnii* Kellogg (= *Q. palmerii* Engelmann). There are 3 or 4 strong secondaries that reach into the prominently toothed margin; the teeth have long, stout prickles; the leaf is wavy or crisped; and the 4th-order venation makes a fine even mesh, much like that of the related *Q. chrysolepis* Liebmann.

The modern *Q. dunnii* is a frequent shrub in the chaparral of interior southern California, chiefly in Riverside and San Diego counties, and occurs also in Arizona. It has a notable disjunct occurrence in the southern Coast Ranges, notably in Peachy Canyon west of Paso Robles, and also in the San Benito Mountains near Idria (Griffin and Tucker, 1976). The Turlock Lake record is apparently a relict from a somewhat earlier time, when a warmer and drier climate enabled a number of taxa whose deriva-tives now have a more southerly distribution to inhabit central California, as reported earlier for the Mulholland (Axelrod, 1944b) and Oakdale (Axelrod, 1944c) floras.

Collection: U.C. Mus. Pal., Paleobot. Ser., Turlock Lake, hypotype nos. 6078-6081; homeotype no. 6082.

Quercus wislizenoides Axelrod
(Plate 11, figs. 1-7)

Quercus wislizenoides Axelrod, Carnegie Inst. Wash. Pub. 553, p. 136, pl. 29, figs. 4-9, 1944; Condit, *ibid.* 553,
  p. 46, pl. 5, figs. 4, 5, 1944; Axelrod, *ibid.* 553, p. 162, pl. 33, figs. 2, 5, 1944; Axelrod, *ibid.* 590, p. 59, pl. 2,
  figs. 11, 12, 1950.

The leaves of this species form the dominant taxon in the flora. They are well preserved, vary from entire to many-toothed, and in a number of cases the fine prickles at the ends of the teeth are preserved owing to the fine matrix in which they are embedded. In all respects these leaves are well matched by those of the living interior live-oak *Q. wislizenii*, which ranges widely in the region, reaching from the oak wood-land belt up into the broadleaved sclerophyll forest. The taxon in the coastal strip that enters the broadleaved sclerophyll Douglas fir and redwood forests appears to be a distinct ecotype as compared with that of the dry digger pine belt. The fossils resemble leaves of the latter, not the former.

Collection: U.C. Mus. Pal., Paleobot. Ser., Turlock Lake, hypotype nos. 6083-6089; homeotype nos. 6090-6110.

Family LAURACEAE
Persea coalingensis (Dorf) Axelrod
(Plate 12, fig. 4)

Persea coalingensis (Dorf) Axelrod, Carnegie Inst. Wash. Pub. 553, p. 132, 1944 (see synonymy and discussion); Condit, *ibid.* 553, p. 79, pl. 16, fig. 6, 1944; Axelrod, *ibid.* 553, p. 201, pl. 38, figs. 1, 3, 1944;
Axelrod, *ibid.* 590, p. 61, pl. 3, figs. 5, 6, 1950; Axelrod, *ibid.* 590, p. 105, pl. 2, fig. 7, 1950; Axelrod, *ibid.*
590, p. 148, pl. 3, fig. 9, 1950.

A single leaf that seems referable to this common Pliocene species has been recovered at each of the two localities in the Upper Mehrten Formation at Turlock Lake. The leaves are incomplete, chiefly because recovery of large specimens is rarely possible, owing to the fractured nature of the claystones in which they occur. The specimens are narrowly lanceolate, and the largest one is somewhat over 8 cm long and 2.3 cm wide. The other specimen, somewhat less complete, is slightly broader and measures 2.4 cm in width and over 9 cm in length.

The fossils show the typical camptodrome secondary venation, and the tertiary veins are also well preserved, displaying a coarse mesh much like leaves of modern *Persea*. The fossil shows relationship to the slender leaves of *P. borbonia* Sprengl of the southeastern United States, and also to those of *P. podadenia* Blake from the mountains of western Mexico.

Collection: U.C. Mus. Pal., Paleobot. Ser., Turlock Lake, hypotype no. 6111; homeo-type nos. 6112-6113.

Umbellularia salicifolia (Lesquereux) Axelrod
(Plate 12, figs. 1-3)

Umbellularia salicifolia (Lesquereux) Axelrod, Carnegie Inst. Wash. Pub. 516, p. 102, pl. 8, fig. 4, 1939 (see
synonymy); Axelrod, *ibid.* 590, p. 62, pl. 3, fig. 2, 1950.

The leaves of California laurel are relatively abundant in the shales at Locality T-20. They are typical of the species in the drier parts of its range, where they tend to be more linear than in the mesic coastal strip, where shape is often oval. All of the specimens are incomplete. The largest one measures 7.8 cm. long and 1.8 cm wide. The tip is relatively blunt, the petiole heavy, the base is broadly acute, the secondaries display the subparallel venation characteristic of the genus, and the tertiaries form a coarse network.

Umbellularia has adapted to very diverse habitats in California, ranging from giant trees (2 m dbh, 35 m tall) that form a savanna on rich alluvial flats in the coastal strip of central California, to an ecotype that is shrubby on serpentine outcrops. These and others were noted and described by Jepson (1910) many years ago.

Collection: U.C. Mus. Pal., Paleobot. Ser., Turlock Lake, hypotype nos. 6114-6116; homeotype nos. 6117-6118.

Family PLATANACEAE
Platanus paucidentata Dorf
(Plate 13, fig. 1)

Platanus paucidentata Dorf, Carnegie Inst. Wash. Pub. 412, p. 94, pl. 10, figs. 4, 9; pl. 11, fig. 1, 1930; Axelrod, *ibid.* 476, p. 174, pl. 5. figs. 4, 5, 1937; Axelrod, *ibid.* 590, p. 62, pl. 4, figs. 1, 8, 1950.

The large leaves of this sycamore are relatively common in the Turlock Lake flora. However, they are recovered chiefly as fragmentary specimens, because the soft clay-stones in which they are preserved are not indurated and readily disintegrate into kernels upon drying, thus fragmenting the specimens.

This sycamore is similar in all respects to the living *P. racemosa* Nuttall, which is scattered along streamways from central California southward into northern Baja California. In central and southern California it reaches up to the margins of forest, and in the south it extends out into the upper desert area. A comparable relation is indicated for its distribution in the Miocene.

Collection: U.C. Mus. Pal., Paleobot. Ser., Turlock Lake, hypotype no. 6119; homeotype nos. 6120-6124.

Prunus turlockensis n. sp.
(Plate 14, fig. 5)

Description: Hard exocarp 11 mm long, 7 mm broad, about 0.5 mm thick; ovate-elliptic in outline, rounded basally, acute distally; surface scarred with sinuous lacunae reminiscent of peach and other prunoids.

Discussion: This individual specimen, figured here at X2, is split longitudinally. It appears to be the hard shell of a species of *Prunus*, but there is insufficient material in herbaria to place it in terms of its relations with any one modern species. Comparison with the figures of diverse *Prunus* species presented in Krussmann (1962) suggests that the closest allies of *P. turlockensis* are in Eurasia, not North America.

Collection: U.C. Mus. Pal., Paleobot. Ser., Turlock Lake, holotype no. 6137.

Family FABACEAE
Amorpha condonii Chaney
(Plate 11, figs. 9-10)

Amorpha condonii Chaney, Carnegie Inst. Wash. Pub. 553, p. 318, pl. 51, figs. 2-4, 1944; Chaney, *ibid.*, p. 348, pl. 63, figs. 1, 2, 1944.

Leaflets of this species, which are similar to those of the living *A. californica* Nuttall, are ovate to oval in outline, broadly rounded above and below, and the camptodrome secondaries are well preserved. The thick petiolule has cross-striations. The modern shrub ranges from Napa and Sonoma counties north of San Francisco Bay southward through the Coast Ranges into southern California, chiefly in the upper part of the woodland belt and reaching locally into forests. It occurs in the central Sierra Nevada in Placer and Eldorado counties, at the lower margin of the forest and in the upper woodland belt.

Collection: U.C. Mus. Pal., Paleobot. Ser., Turlock Lake, hypotype nos. 6125-6126; homeotype nos. 6127-6129.

Family ANACARDIACEAE
Toxicodendron (*Rhus*) *franciscana* Axelrod
(Plate 14, fig. 7)

Rhus franciscana Axelrod, Carnegie Inst. Wash. Pub. 553, p. 141, pl. 30, figs. 9, 11, 1944.

General recognition that *Toxicondendron* is one of several valid segregates of the wastebasket of taxa that have been assigned generally to *Rhus* makes it desirable to rename this species from the Mulholland flora. Its record in the Turlock Lake flora is based on a single ovate lateral leaflet that measures 5.0 cm long, and 3.6 cm wide near the middle of the blade. The margin has remote, blunt teeth, and the secondaries bifurcate and then run into the margin.

The living *T.* (*Rhus*) *diversiloba* is widely distributed in California, and may form a shrub in open areas or a long woody vine clambering 100 ft. (30 m) up into trees.

Collection: U.C. Mus. Pal., Paleobot. Ser., Turlock Lake, hypotype no. 6130.

Family RHAMNACEAE
Ceanothus tuolumnensis n. sp.
(Plate 14, fig. 3)

Description: Basal two-thirds of a distinctive leaf, blade elliptic, base rounded, apex missing, 10 mm long and 6 mm wide; midrib firm and straight, with a prominent basal pair of primaries reaching well toward the apex; tertiary venation obscured; margin with prominent glandular teeth.

Discussion: This leaf seems sufficiently distinctive so that it merits description as a species even though it is not a complete specimen. The leaf is closely matched by those of the modern *C. sorediatus* Hooker and Arnott, a common shrub in the Coast Ranges of California, reaching from Humboldt County southward into the northern part of southern California, as in the Santa Monica Mountains. It often forms a component of the chaparral, and also enters into the woodland zone in the Coast Ranges.

None of the other described fossil species of the subgenus Euceanothus resembles this material.

Collection: U.C. Mus. Pal., Paleobot. Ser., Turlock Lake, holotype no. 6132.

Ceanothus turlockensis n. sp.
(Plate 14, fig. 2)

Description: Leaf oblong-elliptic, narrowed at base, blunt above, petiole missing; midrib firm and straight; blade 17 mm long and 8 mm wide at middle; alternate secondaries diverging at medium angles, complexly looping within the margin, the basal pair more prominent but not forming subprimaries; complexly irregular tertiaries; margin entire.

Discussion: This specimen is similar in all respects to leaves of the living *C. palmerii* Trelease, a common shrub in the lower part of the forest belt in southern California, and occurring also in the central Sierra Nevada in Amador and Eldorado counties at the lower margin of forest and in the upper woodland belt.

No previously described species resembles this taxon, which belongs to the subgenus

Euceanothus. Among those that have been described are *C. edensis* Axelrod from the Pliocene Mt. Eden and Sonoma floras, which is allied to *C. divaricatus* and differs from *C. turlockensis* in its ovate shape and in its prominent basal primaries; *C. leitchii* Axelrod, which resembles the living *C. velutinus*, is much larger than *C. turlockensis*, and has a glandular-serrate margin as well as prominent primaries; *C. prespinosus* Axelrod, which has a much larger, oval leaf; and *C. chaneyii* Dorf, which is also larger and has prominent basal primaries.

Regarding fossil species of the section Cerastes, most of them have been referred to the fossil *C. precuneatus* because they are more nearly like the normal variation of that species than any other. However, Nobs (1962) has expressed the opinion that most of them are more closely allied to *C. megacarpus* and *C. insularis* than to *C. cuneatus* (or *greggii*). Together with the assistance of Howard Schorn, I have now (May 1977) reexamined all of the specimens of *C. precuneatus* discussed by Nobs, using the large annotated collection of species in the Berkeley Herbarium for comparison. We readily concurred that *C. precuneatus* Axelrod, as described from several fossil floras ranging in age from Middle Miocene (Tehachapi flora) into the Pliocene, are in fact most nearly allied to *C. cuneatus*, and that any presumed affinities with *C. megacarpus* or *insularis* find no basis in fact.

As pointed out by Nobs (1962, p. 20), two of the specimens referred to *C. precuneatus* are not *Ceanothus*. The specimen from Remington Hill (Condit, 1944a) is *Forestiera*, and is here transferred to *F. buchananensis* Condit from the Table Mountain flora (Condit, 1944b). The Oakdale specimen of *C. precuneatus* (Axelrod, 1944c) is only a scrap, and since it cannot be identified with reasonable certainty, it is here rejected as a valid record.

Collection: U.C. Mus. Pal., Paleobot. Ser., Turlock Lake, holotype no. 6131.

Rhamnus moragensis Axelrod
(Plate 14, fig. 1)

Rhamnus moragensis Axelrod, Carnegie Inst. Wash. Pub. 553, p. 143, pl. 30, fig. 7, 1944.

A single, well-preserved leaf impression in the flora which is broadly ovate in shape measures 1.9 cm. long and 1.8 cm. wide. The leaf was clearly thick and coriaceous, there are 8 alternate firm camptodrome secondaries connected evidently by a cross-percurrent venation, but the enclosed finer dense mesh cannot be clearly seen. The petiole evidently was thick and short, the margin has numerous sharply pointed teeth, and the firm midrib evidently ended in a thick petiole. Among living species, the leaves of *R. ilicifolia* Kellogg, which is common in the woodland and chaparral of central and southern California, and occurs also in similar vegetation in central Arizona, are not separable from the fossil. *R. ilicifolia* has been considered by some authorities to be a variety of *R. crocea*, but there is little evidence to support such a disposition for the taxon. They often occur sympatrically in California, and it is there that the marked differences become apparent. *R. ilicifolia* is a large shrub to small tree (up to 10 m) with large leaves, whereas *crocea* is a low, intricately branched shrub with small leaves. Both are members of sect. Crocei, which includes other taxa in the southwestern parts of North America, notably *R. insula* and *R. pirifolia*, all of which in my opinion represent the evergreen members of a genus distinct from *Rhamnus*.

Collection: U.C. Mus. Pal., Paleobot. Ser., Turlock Lake, hypotype no. 6133.

Rhamnus precalifornica Axelrod
(Plate 14, fig. 4)

Rhamnus precalifornica Axelrod, Carnegie Inst. Wash. Pub. 516, p. 122, pl. 11, figs. 1, 2, 1939; Axelrod, *ibid.* 590, p. 155, pl. 3, figs. 14, 15, 1950.

The record of this species in the Turlock Lake flora is based on an elliptic leaf impression that exceeded 2.5 cm in length. The thick petiole is 2 mm long, the subparallel alternate camptodrome secondaries are well preserved, as is the cross-percurrent tertiary venation and the finely serrate margin. The specimen is much like leaves of the living *R. californica* Eschscholtz, a widely distributed shrub from the margins of the desert well up into the mixed conifer forest belt.

Collection: U.C. Mus. Pal., Paleobot. Ser., Turlock Lake, hypotype no. 6134.

Family ERICACEAE
Arbutus matthesii Chaney
(Plate 14, fig. 8)

Arbutus matthesii Chaney, Carnegie Inst. Wash. Pub. 346, p. 131, pl. 20, figs. 1, 5 only, 1927; Condit, *ibid.* 553, p. 87, pl. 20, fig. 5, 1944.

The partial specimen illustrated for this flora reveals all the typical venation of *Arbutus* leaves, notably the numerous, thin, wavering secondaries that diverge at medium angles and bifurcate in their outward course. The margin is entire, the texture firm, and the midrib is strong.

Comparison with leaves of *A. menziesii* Pursh shows that they are closely similar. This tree is found locally in the central Sierra Nevada near the lower edge of forest, and increases in abundance from Yuba County northward, but is more prominent in the vegetation of the Coast Ranges. There it contributes to mixed evergreen forest and to the bordering conifer forests as well.

Collection: U.C. Mus. Pal., Paleobot. Ser., Turlock Lake, hypotype no. 6135.

Family OLEACEAE
Forestiera buchananensis Condit
(Plate 14, fig. 6)

Forestiera buchananensis Condit, Carnegie Inst. Wash. Pub. 553, p. 88, pl. 20, fig. 2, 1944.
Ceanothus precuneatus Axelrod. Condit, *ibid.* 553, p. 53, 1944.

The record of this species in the Turlock Lake flora is based on a single oblong-lanceolate leaf impression. The secondary venation is weakly developed and soon approaches the calibre of tertiaries, and these are quite irregular in pattern. The margin is well preserved and shows an occasional bluntish tooth. In all respects, the specimen seems similar to leaves of the modern *F. neomexicana* Gray, a small tree or shrub widely distributed along watercourses in the Southwest, and reaching northward in the inner Coast Ranges of California into the Diablo Range.

Comparison of the leaf of *Ceanothus precuneatus* recorded by Condit at Remington Hill with leaves of *Forestiera* shows that they are inseparable. Examination of the fossil under diffuse light reveals the venation and the serrate margin to best advantage.

Collection: U.C. Mus. Pal., Paleobot. Ser., Turlock Lake, hypotype no. 6136.

REFERENCES CITED

Arkley, R.J.
  1962 The geology, geomorphology and soils of the San Joaquin Valley in the vicinity of the Merced River, California. Calif. Div. Mines and Geology Bull. 182: 25-32.

Axelrod, D.I.
  1944a The Black Hawk flora. Carnegie Inst. Wash. Pub. 553: 91-102.
  1944b The Mulholland flora. Carnegie Inst. Wash. Pub. 553: 103-146.
  1944c The Oakdale flora. Carnegie Inst. Wash. Pub. 553: 147-166.
  1944d The Sonoma flora. Carnegie Inst. Wash. Pub. 553: 167-206.
  1944e The Pliocene sequence in central California. Carnegie Inst. Wash. Pub. 553: 207-224.
  1948 Climate and evolution in western North America during Middle Pliocene time. Evolution 2: 127-144.
  1950 A Sonoma florule from Napa, California. Carnegie Inst. Wash. Pub. 590: 23-71.
  1956 Mio-Pliocene floras from west-central Nevada. Univ. Calif. Pub. Geol. Sci. 33: 1-316.
  1957a Late Tertiary floras and the Sierra Nevadan uplift. Geol. Soc. Amer. Bull. 68: 19-46.
  1957b Paleoclimate as a measure of isostasy. Amer. Jour. Sci. 255: 690-696.
  1962 A Pliocene *Sequoiadendron* forest from western Nevada. Univ. Calif. Pub. Geol. Sci. 39: 195-268.
  1971 Fossil plants from the San Francisco Bay region. *In* Geologic Guide to the Northern Coast Ranges, Point Reyes Region, California, pp. 74-86. Geol. Soc. Sacramento, Annual Field Trip Guidebook.
  1975 Evolution and biogeography of Madrean-Tethyan sclerophyll vegetation. Missouri Bot. Garden Ann. 62: 280-334.
  1976 Evolution of the Santa Lucia fir (*Abies bracteata*) ecosystem. Missouri Bot. Garden Ann. 63: 21-41.

Blackwelder, E.
  1931 Pleistocene glaciation in the Sierra Nevada and Basin Ranges. Geol. Soc. Amer. Bull. 42: 865-922.

Casteel, R.W., and J.H. Hutchinson
  1973 *Orthodon* (Actinopterygii, Cyprinidae) from the Pliocene and Pleistocene of California. Copeia 2: 358-361.

Cavender, T.M., and R.R. Miller
  1972 *Smilodonichthys rastrosus*, a new Pliocene salmonid fish from western United States. Univ. Oregon Mus. Nat. Hist. Bull. 18: 1-44.

Chaney, R.W.
  1938 The Deschutes flora of eastern Oregon. Carnegie Inst. Wash. Pub. 476: 187-216.

Chaney, R.W., and D.I. Axelrod
  1959 Miocene floras of the Columbia Plateau. Carnegie Inst. Wash. Pub. 617. 229 pp.

Chaney, R.W., and M.K. Elias
  1936 Late Tertiary floras from the High Plains. Carnegie Inst. Wash. Pub. 476: 1-46.

Christensen, M.N.
  1966 Late Cenozoic crustal movements in the Sierra Nevada of California. Geol. Soc. Amer. Bull. 77: 163-182.

Condit, C.
  1938 The San Pablo flora of west central California. Carnegie Inst. Wash. Pub. 476: 217-268.
  1944a The Remington Hill flora (California). Carnegie Inst. Wash. Pub. 553: 21-55.
  1944b The Table Mountain flora (California). Carnegie Inst. Wash. Pub. 553: 57-90.

Curry, R.R.
  1966 Glaciation about 3,000,000 years ago in the Sierra Nevada. Science 154: 770-771.

1968 Quaternary climatic and glacial history of the Sierra Nevada, California. Ph.D. thesis, Univ. California, Berkeley.

Dalrymple, G. B.

1963 Potassium-argon dates and the Cenozoic chronology of the Sierra Nevada, California. Ph.D. thesis, Univ. California, Berkeley. 93 pp.

1964 Cenozoic chronology of the Sierra Nevada, California. Univ. Calif. Pub. Geol. Sci. 47. 41 pp.

Davis, S. N., and F. R. Hall

1959 Water quality of eastern Stanislaus and northern Merced counties, California. Stanford Univ. Pubs. Geol. Sci. 6(1): 1-112.

De Laveaga, M.

1952 Oilfields in Central San Joaquin Valley province. *In* AAPG-SEPM-SEG Guidebook Field Trip Routes, pp. 99-103. Joint Ann. Mtg. Am. Assoc. Petrol. Geologists, Soc. Econ. Paleontologists and Mineralogists, Soc. Exploration Geophysicists.

Dibblee, T. W., Jr., and G. B. Oakeshott

1953 White Wolf fault in relation to the geology of the southeastern margin of the San Joaquin Valley, California. Geol. Soc. Amer. Bull. 64 (12), pt. 2: 1502-1503.

Durham, J. W.

1950 Cenozoic marine climates of the Pacific Coast. Geol. Soc. Amer. Bull. 61: 1243-1264.

Durrell, C. D.

1944 Andesite breccia dikes near Blairsden, California. Geol. Soc. Amer. Bull. 55: 255-272.

Elias, M. K.

1942 Tertiary prairie grasses and other herbs from the High Plains. Geol. Soc. Amer. Spec. Papers 41: 1-176.

Evernden, J. F., D. E. Savage, G. H. Curtis, and G. T. James

1964 Potassium-argon dates and the Cenozoic mammal chronology of North America. Amer. Jour. Sci. 262: 145-198.

Gilluly, J., A. C. Waters, and A. O. Woodford

1968 Principles of Geology, 3rd ed. San Francisco: W. H. Freeman. 687 pp.

Goldman, H. B.

1964 Geology of the Tertiary fluvial deposits in the vicinity of Mokelumne Hill, Calaveras County, California. M.A. thesis, Univ. California, Los Angeles. 56 pp.

Griffin, J. R., and W. B. Critchfield

1972 The distribution of forest trees in California. USDA Forest Service Research Paper PSW 82: 1-114.

Griffin, J. R., and J. M. Tucker

1976 Range extension for *Quercus dunnii* in central California. Madroño 23 (5): 295.

Guacci, G., and C. W. Purcell

1978 Evidence for Middle Pleistocene and possible Holocene faulting on the Pond-Poso Creek fault, southern San Joaquin Valley, California. Geol. Soc. Amer., Abstracts with Programs 10 (3): 108.

Hackel, O.

1966 Summary of the geology of the Great Valley. Calif. Div. Mines and Geol. Bull. 190: 215-238.

Hsu, K. J., and 9 co-authors

1977 History of the Mediterranean salinity crisis. Nature 267: 399-403.

Jennings, C. W., and R. G. Strong

1959 Geologic Map of California, Santa Cruz Sheet. Calif. Div. Mines and Geology.

Jepson, W.L.
  1910   The Silva of California. Mem. Univ. California, vol. 2. 480 pp.
Kaneps, A.G.
  1979   Gulf Stream velocity fluctuations during the Cenozoic. Science 204: 297-301.
Kennett, J.P., N.J. Shackleton, S.V. Margolis, et al.
  1979   Late Cenozoic oxygen and carbon isotopic history and volcanic ash stratigraphy: DSDP site 284, South Pacific. Amer. Jour. Sci. 279: 52-69.
Krussmann, G.
  1962   Handbuch der Laubgeholze, vol. 2. Berlin: Paul Parley. 608 pp.
Lamotte, R.S.
  1936   The Upper Cedarville flora of northwestern Nevada and adjacent California. Carnegie Inst. Wash. Pub. 455: 57-142.
Lindgren, W.
  1911   The Tertiary gravels of the Sierra Nevada of California. U.S. Geol. Surv. Prof. Paper 73. 226 pp.
Little, E.L., Jr., and W.B. Critchfield
  1969   Subdivisions of the genus *Pinus* (Pines). U.S. Dept. Agric., Forest Service, Misc. Pub. 1144. 51 pp.
MacGinitie, H.D.
  1953   Fossil plants of the Florissant beds, Colorado. Carnegie Inst. Wash. Pub. 599. 188 pp.
  1962   The Kilgore flora: a Late Miocene flora from Northern Nebraska. Univ. Calif. Pub. Geol. Sci. 35: 67-158.
Mannion, L.E.
  1960   Geology of the La Grange quadrangle, California. Ph.D. thesis, Stanford Univ.
Marchand, D.E.
  1977   The Cenozoic history of the San Joaquin valley and adjacent Sierra Nevada as inferred from the geology and soils of the eastern San Joaquin valley. *In* M.E. Singer (ed.), Soil Development, Geomorphology, and Cenozoic History of the Northeastern San Joaquin Valley and Adjacent Areas, California, pp. 39-50. Guidebook for Joint Field Session of Amer. Soc. Agronomy, Soil Science Soc. Amer. and Geol. Soc. Amer. Univ. California, Davis, Dept. Land, Air, and Water Resources.
Murphy, G.I.
  1950   The life history of the greaser Blackfish (*Orthodon microlepidotus*) of Clear Lake, Lake County, California. Calif. Fish and Game Bull. 36: 119-133.
Noble, D.C., D.B. Slemmons, M.J. Korringa, et al.
  1974   Eureka Valley tuff, east-central California and adjacent Nevada. Geology 2: 139-142.
Nobs, M.A.
  1962   Experimental studies on species relationships in *Ceanothus*. Carnegie Inst. Wash. Pub. 623: 1-94.
Page, W.D., F.H. Swan, III, N. Biggar, et al.
  1978   Evaluation of Quaternary faulting in colluvium and buried paleosols, western Sierran foothills, California. Geol. Sci. Amer., Abstracts with Programs 10 (3): 141.
Piper, A.M., H.S. Gale, H.E. Thomas, and R.W. Robinson
  1939   Geology and ground-water hydrology of the Mokelumne area, California. U.S. Geol. Surv. Water-Supply Paper 780. 230 pp.
Putnam, W.C.
  1962   Late Cenozoic geology of McGee Mountain, Mono County, California. Univ. Calif. Pub. Geol. Sci. 40: 181-218.

Reed, R. D.
    1933   The Geology of California. Amer. Assoc. Petrol. Geol., Tulsa, Okla. 355 pp.
Rogers, T. H.
    1966   Geologic Map of California, San Jose Sheet. Calif. Div. Mines and Geology.
Shackelton, N. J., and J. P. Kennett
    1975   Paleotemperature history of the Cenozoic and the initiation of Antarctic glaciation: oxygen and carbon isotope analyses in D.S.D.P. sites 277, 279, and 281. *In* J.P. Kennett, R. E. Houtz et al. (eds.), Initial Reports of the D.S.D.P. 29, pp. 743-755.
Shackelton, N. J., and N. D. Opdyke
    1976   Oxygen-isotope and paleotemperature stratigraphy of Pacific core V28-239, Late Pliocene to Latest Pleistocene. Geol. Soc. Amer. Mem. 449: 449-464.
    1977   Oxygen isotope and paleomagnetic evidence for early Northern Hemisphere glaciation. Nature 270: 216-219.
Slemmons, D. B.
    1966   Cenozoic volcanism of the central Sierra Nevada, California. Calif. Div. Mines and Geol. Bull. 190: 199-208.
Smith, L. A.
    1977   Messinian event. Geotimes 22 (3): 20-23.
Stirton, R. A., and H. F. Goeriz
    1942   Fossil vertebrates from the superjacent deposits near Knight's Ferry, California. Univ. Calif. Pub. Geol. Sci. 26: 447-472.
Turner, H. W.
    1894   Description of the Jackson quadrangle (Sierra Nevada, California). U.S. Geol. Surv. Geol. Atlas, Folio 11. 6 pp.
Turner, H. W., and F. L. Ransome
    1897   Description of the Sonora quadrangle (California). U.S. Geol. Surv. Geol. Atlas, Folio 41. 5 pp.
    1898   Description of the Big Trees quadrangle (California). U.S. Geol. Surv. Geol. Atlas, Folio 51. 8 pp.
Vanderhoof, V. L.
    1933   A skull of *Pliohippus tantalus* from the later Tertiary of the Sierran foothills of California. Univ. Calif. Pub. Geol. Sci. 23: 183-194.
Wagner, H.
    1976   A new species of *Pliotaxidea* (Mustelidae: Carnivora) from California. Jour. Paleo. 50: 107-127.
Wahrhaftig, C.
    1962   Geomorphology of the Yosemite Valley region, California. Calif. Div. Mines and Geol. Bull. 182: 33-46.
Woodring, W. P., R. Stewart, and R. W. Richards
    1940   Geology of the Kettleman Hills Oil Field, California. U.S. Geol. Surv. Prof. Paper 195. 170 pp.

Chapter III Plates

PLATE 8

Fossil locality and modern vegetation allied to the Turlock Lake flora

FIG. 1. Looking east across Turlock Lake to the Sierra Nevada, in the distance, which is capped by snow in late spring. Scattered oak woodland-grass ("savanna") in the middle distance reaches toward the dominant grassland of the Turlock Lake area. The flora was recovered by Dennis Garber from islands in Turlock Lake.

FIG. 2. Looking west into the floodplain vegetation of Atascadero Creek in the southern Santa Lucia Mountains. *Platanus*, *Populus* and *Salix* are in the watercourse. The oak woodland-grassland on the nearby slopes includes *Quercus agrifolia*, *Q. douglasii*, *Q. lobata* and *Q. wislizenii*. Mixed evergreen forest and chaparral cover the hills a few hundred meters west, reaching up to the summit level of the range.

PLATE 9
Turlock Lake fossils

FIGS. 1, 2. *Pinus sturgisii* Cockerell. Hypotype nos. 6037 and 6036.

FIG. 3. *Typha lesquereuxii* Cockerell. Hypotype no. 6041.

FIG. 4. *Juncus* sp. Axelrod. no. 6039.

FIG. 5. *Smilax remingtonii* Axelrod. Hypotype no. 6040.

FIG. 6. *Cyperus* sp. Axelrod. no. 6038.

FIG. 7. *Salix edenensis* Axelrod. Hypotype no. 6069.

FIGS. 8, 9. *Salix hesperia* (Knowlton) Condit. Hypotype nos. 6061 and 6062.

PLATE 10
Turlock Lake fossils
FIGS. 1-3. *Populus garberii* Axelrod. Hypotype nos. 6043, 6044, 6042.
FIG. 4. *Salix laevigatoides* Axelrod. Hypotype no. 6070.

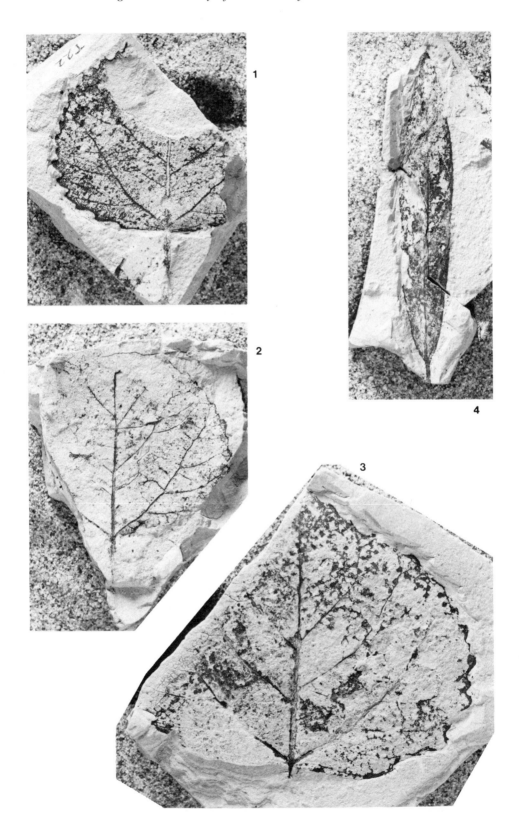

PLATE 11
Turlock Lake fossils

FIGS. 1-7. *Quercus wislizenoides* Axelrod. Hypotype nos. 6083-6089.

FIG. 8. *Quercus douglasoides* Axelrod. Hypotype no. 6075.

FIGS. 9, 10. *Amorpha condonii* Chaney. Hypotype nos. 6125, 6126.

FIG. 11. *Quercus dispersa* (Lesquereux) Axelrod. Hypotype no. 6073.

PLATE 12
Turlock Lake fossils
FIGS. 1-3. *Umbellularia salicifolia* (Lesquereux) Axelrod. Hypotype nos. 6114-6116.
FIG. 4. *Persea coalingensis* (Dorf) Axelrod. Hypotype no. 6111.
FIGS. 5-8. *Quercus pliopalmeri* Axelrod. Hypotype nos. 6078-6081.

PLATE 13
Turlock Lake fossils
Platanus paucidentata Dorf. Hypotype no. 6119.

PLATE 14
Turlock Lake fossils

FIG. 1. *Rhamnus moragensis* Axelrod. Hypotype no. 6133.

FIG. 2. *Ceanothus turlockensis* Axelrod. Holotype no. 6131.

FIG. 3. *Ceanothus tuolumnensis* Axelrod. Holotype no. 6132.

FIG. 4. *Rhamnus precalifornica* Axelrod. Hypotype no. 6134.

FIG. 5. *Prunus turlockensis* Axelrod. Holotype no. 6137. X-2.

FIG. 6. *Forestiera buchananensis* Condit. Hypotype no. 6136.

FIG. 7. *Toxicodendron (Rhus) franciscana* Axelrod. Hypotype no. 6130.

FIG. 8. *Arbutus matthesii* Chaney. Hypotype no. 6135.

IV

THE BROKEN HILL FLORA
FROM KINGS COUNTY

Chapter IV Contents

FIGURES

TABLES

PLATES

INTRODUCTION

One of the thickest and most completely studied Late Tertiary sections in California is in the Kettleman Hills (Woodring, Stewart, and Richards, 1940). Situated in the lee of the inner south Coast Ranges southeast of Coalinga, the exposed Pliocene and Lower Pleistocene rocks are fully 6,000 ft. (1,825 m) thick, and the section increases to over 8,000 ft. (2,430 m) 4 miles (2.5 km) west in the Kreyenhagen Hills (Stewart, 1946). The general rarity of remains of land life in these rocks reflect their predominantly marine nature, as shown by numerous beds filled with marine mollusks. Occasional fragmentary mammalian remains have been recovered, but plants are quite rare.

A few leaf impressions collected at two sites by Woodring and his associates (1940) in the Kettleman Hills were identified by R. W. Brown. The locality from the *Siphonalia* zone in the middle of the Etchegoin Formation yielded 6 taxa, distributed in *Alnus*, *Fraxinus*, *Persea* (their *Salix*), *Platanus*, *Populus*, and *Quercus*, and the *Pecten* zone in the middle of the San Joaquin Formation yielded 3 species, distributed in *Platanus*, *Persea* (listed as *Salix*) and *? Umbellularia*. Although the collections were not analyzed or retained, it is evident that they largely represent members of floodplain vegetation marginal to the marine embayment. The species are similar to those in the small flora described by Dorf (1930) from the *Pecten* zone in the Kreyenhagen Hills situated about 8 miles (13 km) northwest. Composed of *Alnus*, *Garrya*, *Persea*, *Platanus*, *Populus*, and *Quercus*, they also represent members of floodplain vegetation chiefly, with oaks on nearby well-drained substrates. Woodring and his associates also refer in their descriptive sections to plants at several other sites in the Kettleman Hills. All of them have been examined, but the plant fossils are so rare that they add little to that already on record; they represent the same taxa, with species of *Platanus*, *Populus*, and *Persea* present at most of the sites.

In view of the rarity of fossil plants in the Pliocene section, the recovery of a flora of 22 species from the basal Cascajo Member of the San Joaquin Formation is particularly welcome. It provides us with a more complete picture of land environments in the inner south Coast Range area during the Pliocene. The site was discovered by Ralph Stewart during his field investigations after the principal report on the Kettleman Hills (Woodring, Stewart, and Richards, 1940) was completed. The locality is on the crest of Broken Hill, at the south end of North Dome. As shown on the Los Viejos Hills Quadrangle (U.S. Geol. Surv., scale 1:31,680, ed. 1934), Broken Hill is just west of the middle of sec. 36, T. 22 S, R. 18 E, about 4 miles (6.5 km) west of Kettleman City at an elevation of 715 ft. (218 m).

PRESENT PHYSICAL SETTING

Kettleman Hills is a 30-mile-long (38 km), 5-mile-wide (8 km), slightly sinuous anticline that trends southeasterly from near Coalinga. Forming a low ridge that rises 500-600 ft. (150-180 m) above the adjacent plain, it is structurally an outlier of the Coast

Ranges and is separated from them by the grassy Kettleman Plain, which is 2-3 miles (3-4.5 km) wide. The inner Coast Ranges to the west are rugged, with summit levels in this area rising to 3,000-4,000 ft. (900-1,200 m). To the east, Kettleman Hills is bordered by the level San Joaquin Valley, which stretches for 60 miles (96 km) to the front of the Sierra Nevada. The region is semi-arid, with 6 in. (15 cm) annual rainfall coming chiefly during the cooler half of the year, and mostly from November into April. In midsummer the daily high temperatures regularly run over 100°F (38°C). Winters are mild, with occasional frost during the period from December into February. Light snow may intermittently dust the higher Coast Ranges during winter storms. Mean annual temperature is 65°F (18.3°C) at Kettleman Station, 9 miles (14.5 km) north of Broken Hill at an elevation of 508 ft. (155 m). The range of mean temperature, measured as the difference between the mean temperatures of the warmest and coldest months, is 35°F (19.4°C). Warmth of climate is *W* 59.3°F (*W* 15.2°C), or 217 days with a mean temperature at or exceeding that level (see Bailey, 1960; 1964). The equability rating is *M* 48, graded on a scale of 100. By contrast, stations in the coastal strip 100 miles west have ratings of *M* 73 (Monterey), *M* 76 (Pismo Beach), and *M* 79 (Pt. Piedras Blancas).

As a result of low rainfall in the lee of the Coast Ranges, the Kettleman Hills and the adjacent valley region are covered with grassland, and saltbush (*Atriplex*) occurs as a widely scattered shrub. As rainfall gradually increases at higher levels in the Coast Ranges to the west, woodland and chaparral communities appear, and riparian taxa—chiefly *Platanus racemosa, Populus fremontii, Salix exigua, S. lasiolepis,* and other willows—line the drainageways. The lower parts of the Coast Ranges are covered with a *Juniperus californica-Quercus douglasii-Q. dumosa* woodland savanna which gives way at higher elevations to digger pine woodland, composed of *Pinus sabiniana, Quercus agrifolia, Q. douglasii, Q. lobata,* and *Q. wislizenii.* Among the commoner shrubs scattered in the woodland that locally form a dense chaparral are *Adenostema fasciculatum, Arctostaphylos glauca, Ceanothus cuneatus, Cercocarpus montanus, Fraxinus dipetala, Heteromeles arbutifolia, Prunus ilicifolia,* and *Quercus dumosa.* Woodland and chaparral attain best development at levels about 1,500-1,800 ft. (455-548 m), where precipitation is 20 in. (50 cm) or more.

GEOLOGIC OCCURRENCE

The exposed section in the Kettleman Hills consists of Pliocene marine and Early Pleistocene nonmarine rocks (Woodring, Stewart, and Richards, 1940). In descending order, these include the nonmarine Tulare Formation (1,700-3,500 ft.; 517-1,080 m), which is composed chiefly of sandstone and conglomerate, with finer-grained rocks, in which fresh-water mollusks are common, toward the base. It rests conformably on the San Joaquin Formation (1,200-1,800 ft.; 365-548 m), consisting of siltstone, claystone, and pebble conglomerate, with many of the coarser strata rich in marine mollusks. The Etchegoin Formation, which underlies the San Joaquin with local unconformity, consists principally of marine sandstone, conglomerate, and siltstone. It increases in thickness from about 700 ft. (213 m) in the north to 1,800 ft (548 m) on South Dome.

The San Joaquin Formation was divided into an upper and lower part by Woodring et al. (1940, p. 28), and subdivided on the basis of faunal zones, which were mapped. As presented by them, the sequence on the east flank of the North Dome is as follows:

(Local unconformity)
Etchegoin Formation

Evidence for a local unconformity at the base of the Cascajo Member is well illustrated in their figures 6, 7, 8 and, as they suggest, appears to reflect movements of the Coast Ranges to the west, the source area of the pebbles in the Cascajo Member.

The Cascajo Member at Broken Hill (Plate 15, Fig. 1) is chiefly a coarse, cross-bedded, blue-gray sandstone, together with thin interbeds of silty sandstone and siltstone, all of which contain fossil leaves scattered through about 6 ft. (2 m) of section. Some of the large sandstone plates have leaves of sycamore, poplar, and avocado plastered on them, but are difficult to collect because the strata are poorly indurated. It is noteworthy that certain strata (e.g., specimens 6267 and 6268) have layers of numerous leaves and comminuted plant material that resemble accumulations similar to those that collect in quiet water along streams today. The abundance of leaves suggests that in this particular area the Cascajo probably was land-laid, though this is not easy to demonstrate. Also in favor of this suggestion, however, is the apparent absence of marine fossils in this part of the Cascajo, and they are rare elsewhere in this member (Woodring et al., 1940, p. 53). Clearly, if the leaves were deposited in an open sea well offshore, the question arises as to how they would become concentrated by currents in a local area and through a number of feet of section. In this regard, Woodring and his associates have noted that marine fossils, which occur infrequently in the Cascajo, usually are worn, and many appear to have been transported prior to burial. It seems likely that they are the result of penecontemporaneous erosion, being reworked from other parts of the section. Since the Cascajo appears to represent a blanket sheet deposit resulting from uplift in the Coast Ranges, local deltas may have extended eastward into the shallow San Joaquin Sea, supporting vegetation of which the Broken Hills flora is a sample. Delta floodplains of this sort were of only brief duration, however, for the succeeding members of the San Joaquin Formation certainly are of marine origin, containing numerous molluscan horizons.

COMPOSITION OF THE BROKEN HILL FLORA

As presently known, the Broken Hill flora from the basal part of the San Joaquin Formation is represented by 22 species. It is noteworthy that conifers are not present. The taxa are all dicots, distributed among 8 families, each of which is represented by 1 or 2 genera. Three genera, *Populus, Salix,* and *Quercus,* have 3, 4, and 5 species respectively. Fruits or seeds were not encountered in this deposit. As judged from the morphology of the leaves available for study, most of which are well preserved, all of the taxa seem to be indistinguishable from living species.

Systematic List of Species

Salicaceae
  Populus alexanderii Dorf
  Populus coalingensis new species
  Populus garberii Axelrod
  Salix edenensis Axelrod
  Salix hesperia (Knowlton) Condit
  Salix laevigatoides Axelrod
  Salix wildcatensis Axelrod
Betulaceae
  Alnus corrallina Lesquereux
  Fagaceae
  Lithocarpus klamathensis
    (MacGinitie) Axelrod
  Quercus douglasoides Axelrod
  Quercus lakevillensis Dorf
  Quercus pliopalmerii Axelrod
  Quercus remingtonii Condit
  Quercus wislizenoides Axelrod

Ulmaceae
  Celtis kansana Chaney and Elias
  Ulmus affinis Lesquereux
Magnoliaceae
  Magnolia corrallina Chaney and
    Axelrod
Lauraceae
  Persea coalingensis (Dorf) Axelrod
  Umbellularia salicifolia
    (Lesquereux) Axelrod
Platanaceae
  Platanus paucidentata Dorf
Fabaceae
  Amorpha oklahomensis (Berry)
    Axelrod
Sapindaceae
  Sapindus oklahomensis Berry

The general habit of the plants can be judged from that of their nearest living descendants. As listed in Table 11, the flora is composed of 14 trees and 8 shrubs. Two of the tree genera, *Populus* and *Quercus,* are represented by several species, and so is one of the shrubs (*Salix*).

The collection is small, totalling only 238 specimens (Table 12). Nonetheless, it is evident that the abundant plants are those that, as judged from the ecologic occurrences of their nearest living allies, probably lived along the margins of the distributaries of the delta that evidently reached out into the San Joaquin Sea. Noteworthy is the dominance of *Persea,* which at this time was confined chiefly to floodplains where there was ample water. Its occurrence elsewhere in the Pliocene seems similar, as shown by the Anaverde, Mt. Eden, and Sonoma floras (Axelrod, 1950). The closest descendant of the commonest oak is *Quercus wislizenii.* It regularly reaches down stream valleys from moister, higher levels, where it is a regular member of sclerophyll oak woodland vegetation. The leaves of willow, poplar, and sycamore are also prominent, a relation expectable in a sample of riparian-border vegetation. Of notable interest is the abundance of *Q. remingtonii.* It is a hybrid of *Q. wislizenoides* and *Q. deflexiloba* ("*pseudolyrata*"), which are fossil counterparts of *Q. wislizenii* and *Q. kelloggii* that hybridize today to produce *Q. morheus* (see

TABLE 11

Living Species Most Similar to Broken Hill Taxa
(arranged according to their usual life form or habit)

Fossil species	Allied living taxa
Trees (14)	
Alnus corrallina	*A. rhombifolia*
Lithocarpus klamathensis	*L. densiflorus*
Magnolia corrallina	*M. virginiana; grandiflora*
Persea coalingensis	*P. carolinensis; podadenia*
Platanus paucidentata	*P. racemosa; wrightii*
Populus alexanderii	*P. trichocarpa*
Populus coalingensis	*P. euphratica* (section)
Populus garberii	*P. tremula*
Quercus douglasoides	*Q. douglasii*
Quercus lakevillensis	*Q. argifolia*
Quercus remingtonii	*Q. morheus*
Quercus wislizenoides	*Q. wislizenii*
Ulmus affinis	*U. americana*
Umbellularia salicifolia	*U. californica*
Shrubs (8)	
Amorpha oklahomensis	*A. fruticosa*
Celtis kansana	*C. reticulata*
Quercus pliopalmerii	*Q. dunnii*
Salix edenensis	*S. exigua; fluvatilis*
Salix hesperia	*S. lasiandra*
Salix laevigatoides	*S. laevigata*
Salix wildcatensis ·	*S. lasiolepis*
Sapindus oklahomensis	*S. drummondii*

Jepson, 1910, pp. 46-48), a process that had already commenced by the Late Miocene (Condit, 1944). Since the parental oaks grow together today chiefly in the middle and upper parts of the oak woodland belt, the frequent occurrence of *Q. remingtonii* in the flora suggests the nature of the climate that probably typified the area during the Late Pliocene (see below).

PALEOECOLOGY

Setting

Kettleman Hills is situated in the middle of an area that was an inland sea from the Late Cretaceous into the close of the Tertiary. As shown by paleogeographic maps that chart the areas of land and sea (e.g., Reed, 1933, p. 252, fig. 51; Woodring, Stewart, and Richards, 1940, p. 105, fig. 21; Hackel, 1966, p. 235, fig. 9; Galehouse, 1967, fig. 28), the Pliocene seaway reached northwestward across the trend of the present Coast Ranges to enter the Pacific Ocean near Monterey, which was then situated south of its present area, for it lies west of the San Andreas fault. The Diablo uplift to the north and the Temblor uplift to the west and south were important source areas for sediment in the Kettleman

TABLE 12

Numerical Representation of Taxa
in the Broken Hill Flora

Fossil species	Total specimens
Persea coalingensis	45
Platanus paucidentata	41
Populus garberii	35
Quercus wislizenoides	34
Quercus remingtonii	28
Salix laevigatoides	14
Amorpha oklahomensis	8
Salix wildcatensis	5
Quercus pliopalmerii	5
Salix hesperia	4
Quercus douglasoides	3
Populus alexanderii	2
Quercus lakevillensis	2
Salix edenensis	2
Magnolia corrallina	2
Umbellularia salicifolia	2
Alnus corrallina	1
Celtis kansana	1
Lithocarpus klamathensis	1
Populus coalingensis	1
Ulmus affinis	1
Sapindus oklahomensis	1
Total	238

Hills. In addition, transport of sediment from the east is implied by the mineralogy of a number of tuffaceous sandstones in the section. They contain minerals derived from the weathering of hornblende andesite, augite andesite, and hypersthene andesite (Bramlette, in Woodring et al., 1940, p. 50), all of which are exposed in the central Sierra Nevada and northward. The marine beds give way a few miles north of Coalinga to a nonmarine sequence that was filling the Great Valley.

Prior to San Joaquin deposition, the inland sea that occupied the southern part of the present Great Valley had an arm that reached southwestward to the upper part of Salinas Valley near Paso Robles, and thence farther southwest to join the Santa Maria basin. Galehouse (1967) showed that this connection apparently ceased as the La Panza Range was elevated to provide sediment for the Paso Robles Formation from the east during the "Early Pliocene" (now = Late Miocene). During San Joaquin deposition, therefore, the only connection to the Pacific was to the northwest across the inner Coast Ranges and down the San Andreas rift to join the ocean at Monterey Bay. Galehouse estimates that displacement along the rift since San Joaquin deposition amounts to about 25 miles (40 km). As suggested by Woodring and his associates (1940, p. 99), the change from saline to brackish and fresh water during late Etchegoin and early San

TABLE 13

Broken Hill Taxa
Arranged According to Their Probable Vegetation Preferences
(as inferred from that of similar living species)

Taxa	Floodplain, riparian border	Oak woodland-grass	Mixed evergreen forest
Alnus corrallina	x		
Celtis kansana	x		
Platanus paucidentata	x		
Populus alexanderii	x		
Populus coalingensis	x		
Populus garberii	x		
Salix edenensis	x		
Salix hesperia	x		
Salix laevigatoides	x		
Salix wildcatensis	x		
Sapindus oklahomensis	x		
Ulmus affinis	x		
Amorpha oklahomensis	x		
Quercus douglasoides		x	
Quercus lakevillensis		x	
Quercus pliopalmerii		x*	
Quercus remingtonii		x	x
Quercus wislizenoides		x	x
Persea coalingensis	x		x
Magnolia corralina	x		x
Umbellularia salicifolia	x	x	x
Lithocarpus klamathensis			x

*Also contributes to chaparral.

Joaquin deposition probably was due to shoaling, because the rate of deposition exceeded the rate of subsidence of the basin. Offshore bars gradually enclosed estuaries that locally at times were filled with fresh water, and their banks supported vegetation of which the record at Broken Hill appears to be a sample. The final elimination of marine waters from the area at the close of San Joaquin deposition was due to the blocking of the connection with the Pacific Ocean by deformation in the adjoining Coast Ranges.

Vegetation

The plants in the Broken Hill flora are about equally divided between those that regularly have a floodplain or riparian-border occurrence and those that represent woodland and woodland-grass vegetation of the nearby interfluves and slopes. The species of the principal communities that inhabited the area are listed in Table 13, where the taxa are grouped according to the usual environmental preferences of their nearest modern relatives. Nearly all of the woodland taxa descend slopes that border rivers and floodplains and also enter into the makeup of the flora and vegetation there, though chiefly in sites where there is good drainage. Although the riparian-border taxa reach optimum development along stream valleys and floodplains with a high water-table, all of them also occur at seepages and springs on slopes away from the floodplains. However, such occurrences are scattered and are not important in terms of regional vegetation.

To the listing above for woodland, we should note that *Garrya masonii* Dorf, from the slightly younger Kreyenhagen Hills flora to the west (Dorf, 1930), is also a member of oak woodland vegetation and no doubt was in the region when the Broken Hill flora was living. We can thus visualize the shore area and entering rivers supporting floodplain vegetation that was richer than that now in California. Absent from the region today are the genera *Magnolia*, *Persea* (which was dominant), and *Sapindus*, as well as species of *Populus* that are now in floodplain sites in summer-rain climates. The x's in Table 13 that extend the range of *Persea*, *Magnolia*, and *Ulmus* into the floodplain environment imply that this was a compensatory site for them. They were relict in California at this time, and entered the woodland only in its more mesic parts. The bordering slopes were covered with an oak woodland-savanna which in composition was like that still living in the Coast Ranges to the west. There the woodland taxa regularly reach down to interfinger with floodplain species of *Alnus*, *Fraxinus*, *Platanus*, *Populus*, and *Salix* in well-drained sites, as seen in Plate 15, Figure 2, and Plate 8, Figure 2.

Relations of this sort can be seen today in the inner Santa Lucia Mountains, notably in the middle-upper Carmel River Valley, in the upper Nacimiento River drainage near San Antonio Mission, and farther south, in the hills west of Atascadero, Paso Robles, and Santa Margarita. Plants observed in three of these areas that have equivalents in the fossil flora are listed in Table 14. Plate 15, Figure 2, presents a view across the Nacimiento River from the site of the Nacimiento weather station a few miles west of San Antonio Mission, where these taxa contribute to woodland-grass and riparian-border vegetation at the lower margin of mixed evergreen forest and chaparral.

Apart from the listed taxa, a number of additional species recur in all these areas, and contribute importantly to the vegetation, but are not now known from the Broken Hill flora. Notable among these are *Fraxinus oregona* and *Populus fremontii*, both of which are prominent in riparian-border vegetation today. Although *Fraxinus* was recorded by Dorf (1930) from the overlying *Pecten coalingensis* zone in the Kreyenhagen Hills to the west, and Brown (in Woodring et al., 1940, table facing p. 78) lists it in the small collection from the *Siphonalia* zone in the underlying Etchegoin Formation in the Kettleman Hills, those specimens are *Persea*. In addition, important taxa that contribute to the middle and upper part of the mixed evergreen forest in the Santa Lucia Mountains that do not have equivalent species in the Broken Hill flora include *Arbutus menziesii*, *Quercus chrysolepis*, and their frequent associate, *Acer macrophyllum*, which prefers moist sites and stream-border habitats. These occur chiefly at higher, moister levels above the oak woodland-grass belt, interfingering with it as shown on Plate 8, Figure 2. That mixed evergreen forest covered the low inner Coast Ranges to the west of Broken Hill is implied by the rare record (a scrap) of *Lithocarpus* in the flora and by the presence of *Quercus remingtonii*, one of whose parents (*Q.* aff. *kelloggii*) is a common member of that community, as well as the lower conifer forest dominated by *Pinus ponderosa* and *Pseudotsuga menziesii*.

Common shrubs in the Nacimiento area that are not recorded by allied species in the

TABLE 14
Living Species in the Santa Lucia Mountains
that Have Close Fossil Relatives
in the Broken Hill Flora

Modern equivalents of Broken Hill flora	Middle Carmel R. Valley	Upper Nacimiento R.	Atascadero Creek
Populus trichocarpa	x	x	x
Salix lasiandra	x	x	x
Salix exigua	x	x	x
Salix laevigata	—	x	x
Salix lasiolepis	x	x	x
Alnus rhombifolia	x	x	x
Lithocarpus densiflorus	x	x	x
Quercus agrifolia	x	x	x
Quercus douglasoides	x	x	x
Quercus dunnii (= *palmeri*)	—	—	x*
Quercus morheus	—	—	—
Quercus wislizenii	x	x	x
Umbellularia californica	x	x	x
Platanus racemosa	x	x	x

*A typical southern California species that has a relict occurrence near Paso Robles, a few miles north of Atascadero.

Broken Hill flora include *Arctostaphylos* (several species), *Amelanchier pallida, Ceanothus cuneatus, C. sorediatus, Cercocarpus montanus, Fraxinus dipetala, Garrya elliptica* (recorded in the Kreyenhagen Hills flora; Dorf, 1930), *Heteromeles arbutifolia, Laurocerasus* (= *Prunus*) *ilicifolia,* and *Quercus dumosus.* The absence of equivalents of these and other comparable taxa in the fossil flora seems understandable on the basis of edaphic conditions on the Cascajo floodplain. The high water-table would have been unsuited for most of them, because they regularly occur in well-drained sites, often on shallow rocky soil. As judged from the available sample, the Broken Hill flora compares favorably with taxa that now contribute to the middle-upper part of the oak woodland zone, but not within the ecotone with mixed evergreen forest (*Arbutus-Lithocarpus-Quercus chrysolepis*) which lives above it in a wetter, cooler, and more equable climate.

Comparisons of this sort are misleading unless the species that no longer have equivalents in the region are taken into account. A total of 8 species in the Broken Hill flora do not have modern analogues in the Coast Ranges, and 6 of them are no longer in central California. *Amorpha fruticosa* and *Celtis reticulata* occur in southern California bordering the Sonoran Desert, where summer rainfall is more frequent. Both are more prominent in the vegetation in the southwestern to central United States and adjacent Mexico. The other 6 taxa live only in areas with ample rain in summer. The nearest occurrence for *Sapindus* is in the woodland belt in southern Arizona. *Persea* is in northern Mexico in mesic woodland vegetation, and is represented also on the Atlantic and Gulf coast. *Populus euphratica* ranges from central Asia to Africa, from the margins of forest out into drier semidesert areas. *Ulmus* is in the eastern United States, reaching westward up river valleys to interfinger with oak woodland vegetation in the Edwards Plateau of west-central Texas. *Populus tremula*, a close living analogue of *P. garberii*, is wide in the Eurasian region and chiefly of forest occurrence there; the allied *P. grandidentata* of the eastern United States is also a forest species. The rare *Magnolia* in the flora is allied to taxa that are confined chiefly to well-drained, rich floodplains of the Atlantic and Gulf coastal plain. Clearly, all of them would be at home in the Broken Hill flora, provided there was adequate summer rainfall. This raises the problem of the nature of the climate in the region when the flora lived there.

Climate

Inasmuch as the Broken Hill species are scarcely separable from living plants, the conditions under which they live provide an indication of the climate during the Pliocene. The comparison is strengthened because 14 of the species are still associated in plant communities on the inner slopes of the Santa Lucia Mountains. Precipitation at several stations in that area is presented in Table 15. Most of the taxa occur below the middle-upper woodland belt that shows greatest affinity with the fossil flora. This implies a higher rainfall for the fossil flora than that at these sites, or one closer to that at the Nacimiento station (Plate 15, Fig. 2), or a minimum of about 25 in. (64 cm). If it was much higher, then members of broadleaved sclerophyll vegetation (*Acer* cf. *macrophyllum, Arbutus, Lithocarpus, Quercus* aff. *chrysolepis*) would be expected to have a fair representation in the flora. If precipitation was much below 25 in. (64 cm), then there is a problem of accounting for the presence of *Lithocarpus* in the flora, as well as *Quercus remingtonii*. As noted above, one of its parents (*Q.* aff. *kelloggii*) is not found today in areas where rainfall is much lower. Furthermore, several taxa that have equivalents in the flora live in summer-rain areas where rainfall is not much below 25-30 in. (64-76 cm), notably species of *Magnolia, Persea, Populus* (cf. *tremula*), and *Ulmus*. Those that live in regions with summer rain that reach out into drier areas, as *Amorpha* (cf. *fruticosa*), *Celtis*, and *Sapindus*, regularly occur there along drainageways where the high water-table compensates for low precipitation.

Rainfall was present in some amount during summer, as implied by 3 genera (*Magnolia, Persea, Sapindus*) that occur only in such areas, and by several species in the flora whose equivalents (*Amorpha fruticosa, Celtis reticulata, Populus euphratica, P. tremula*) are confined to regions with warm-season precipitation, or at least attain best development there (*Amorpha, Celtis*). Similar floristic relations have been noted earlier for the essentially contemporaneous Sonoma and Napa floras (Axelrod, 1944, 1950), where species of *Castanea, Ilex, Persea, Pterocarya*, and *Ulmus* have been recorded. The amount of summer precipitation at this time is not determinable, though it certainly was not as high as earlier in the Neogene. In attempting to make an estimate, it is essential to emphasize that today precipitation in this area decreases gradually in spring, that the midsummer period is essentially rainless, and that rains commence gradually in autumn (Fig. 14). In this regard, however, note that the stations in southern California (Fig. 14: Campo, Mill Creek) have some summer precipitation, and it increases considerably at higher elevations in the pine-fir forests in the mountains where stations (Big Bear Lake, Cuyamaca) record 1-2 inches (2.5-5 cm) or more in each summer month. It is precisely in these areas that taxa disjunct from the Rocky Mountains have relict

TABLE 15

Climatic Data for Kettleman Hills Area
and for Stations in the Santa Lucia Mountains
Situated near Vegetation Similar to That
in the Broken Hill Flora

Stations and years of record	Annual precipitation	Mean temperature			Annual Range
		Annual	July	January	
Kettleman Hills					
Kettleman Station (20)	5.9 in.	65.0°F	84.3°F	47.2°F	37.1°F
	15.0 cm	18.3°C	29.1°C	8.4°C	20.7°C
Kettleman City (15)	4.6 in.	64.5°F	85.1°F	44.6°F	40.5°F
	11.7 cm	18.1°C	29.5°C	7.0°C	22.5°C
Santa Lucia Mountains					
*‡ Atascadero (13)	18.0 in.	55.4°F	67.7°F	43.7°F	24.0°F
	45.7 cm	13.0°C	19.8°C	6.5°C	13.3°C
* Hastings Reserve (30)	19.8 in.	57.2°F	69.0°F	47.5°F	21.6°F
	50.5 cm	14.0°C	20.6°C	8.6°C	12.0°C
Hunter Liggett A.F.B. (14)	15.3 in.	59.5°F	73.2°F	45.7°F	27.5°F
	38.8 cm	15.3°C	22.9°C	7.2°C	15.3°C
* Nacimiento Dam (18)	18.0 in.	59.4°F	72.1°F	49.1°F	23.0°F
	45.7 cm	15.2°C	22.3°C	9.5°C	12.8°C
Nacimiento River (14)	29.1 in.	58.2°F	70.5°F	46.0°F	24.5°F
	73.6 cm	14.6°C	21.4°C	7.8°C	16.6°C
* Paso Robles (68)	15.0 in.	58.5°F	71.7°F	46.0°F	25.7°F
	38.1 cm	14.7°C	22.1°C	7.8°C	14.3°C
† Santa Margarita (27)	28.0 in.	58.4°F	72.9°F	43.0°F	29.9°F
	71.1 cm	14.7°C	22.7°C	6.1°C	16.5°C

*Stations low in woodland zone have lower precipitation than at higher levels nearby.
†High rainfall reflects its elevation and near-coastal position.
‡Cold air drainage gives a lower mean temperature.

occurrences in southern California, as discussed elsewhere (Axelrod, 1976, p. 32; Raven and Axelrod, 1978, pp. 48, 55 ff.).

My observations suggest that rainfall in California during the Pliocene may have been like that now in the central Mediterranean region. In that area, there is an abrupt increase in rainfall following the dry (but not rainless) summer, and high rainfall persists into late spring, as illustrated for stations along the shores of the north Mediterranean basin (Fig. 15). In this regard, note that Kusadasi, situated on the south coast of Turkey in the sclerophyll belt, has two rainless summer months, yet precipitation increases in September and rises abruptly in October when copious amounts fall there as well as elsewhere in the Mediterranean basin. With an inch or so (2.5-3.0 cm) during each

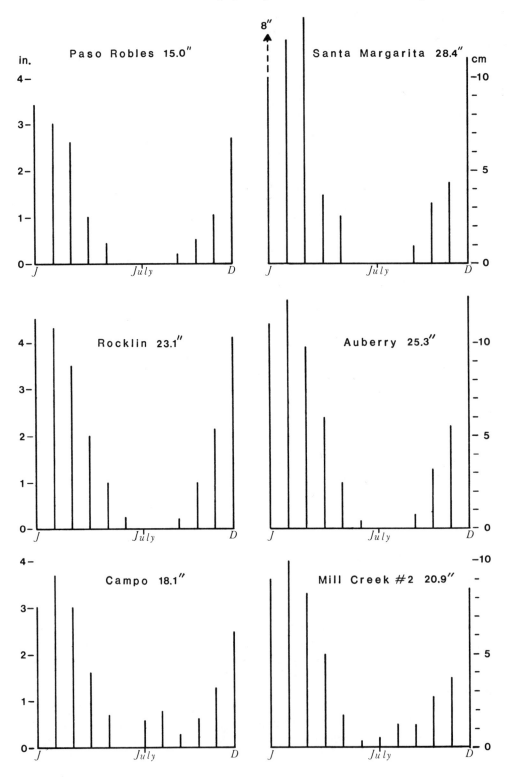

FIG. 14. Distribution of monthly rainfall in sclerophyll vegetation of central and southern California. Note that the long dry summer period is followed by a gradual increase in rainfall in autumn.

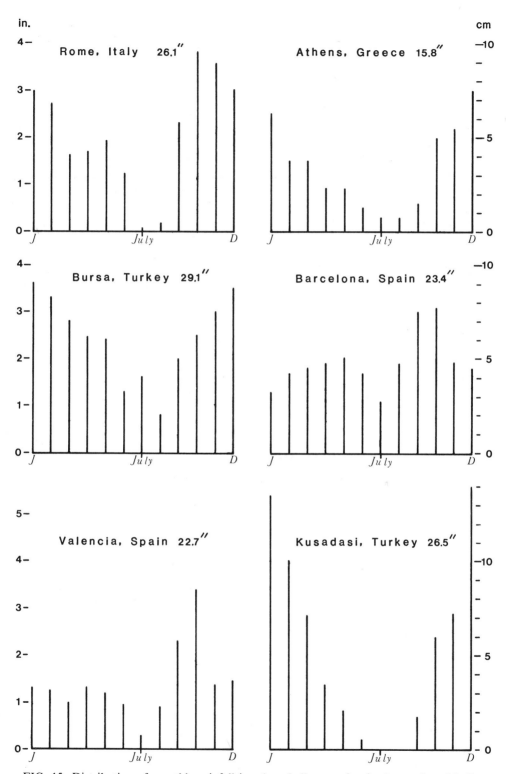

FIG. 15. Distribution of monthly rainfall in sclerophyll vegetation in the northern Mediterranean basin. Compared with California, the summer period is not so dry, rains increase appreciably in September-October, and are also heavier into late spring. Comparable conditions are inferred for the Broken Hill flora.

summer month, and with a rapid increase in rain in September-October, and heavier rain into April-May, drought stress would be greatly alleviated, because temperatures were equable. This could enable summer-rain indicators (*Magnolia, Persea, Sapindus*, etc.) to linger on in the coastal strip.

Analogy with the Mediterranean region is also seen in that it is an inland sea (though larger) like that which covered south-central California in the Pliocene, and it is also warmer than the Atlantic, as was the Etchegoin-San Joaquin sea in relation to the open Pacific. Furthermore, there is light summer rain along the northern Mediterranean shore, where sclerophyll vegetation interfingers with mixed deciduous forest, as in Italy, Greece, and Turkey. Summer rain indicators such as *Betula, Castanea, Tilia, Ulmus*, and others have survived in these areas. Since a similar ecotone existed in California during the Pliocene, and since seas were warmer than those at present, the comparison seems valid. Thus, we may conclude tentatively that light rain in each summer month, a rapid increase in autumn rainfall, and heavier monthly precipitation into late spring, coupled with equable temperature, seem adequate to explain the record as now known.

The data suggest that total precipitation has decreased about 15-18 in. (38-45 cm) over that of the Pliocene. This was due in large part to the elevation of the Coast Range barrier following deposition of the sediments of the San Joaquin and Tulare formations. As a result, oak woodland has been restricted to elevations fully 1,000-1,500 ft. (300-450 m) above the level of the grassland of the San Joaquin Valley, and separated from it by the *Juniperus californica-Quercus douglasii-Q. dumosa-Stenotopsis linearifolius-Yucca whippleyi* community which subsequently invaded the region from the warmer, drier interior, probably following the last glacial (Axelrod, 1966, p. 48). The drier climate over this interior region has also resulted in the elimination of most riparian taxa from drainageways that reach the valley floor. They do not extend away from the foothills of the inner Coast Ranges, except for a few scattered clumps of *Platanus* and *Populus* in the largest stream courses.

The general nature of the temperature change in the region is suggested by comparing thermal conditions in the Kettleman Hills with those in the inner Santa Lucia Mountains (Table 15 and Fig. 16). The data show that the Kettleman Hills area now has a mean temperature at least 6°F (3.3°C) higher than that in the Santa Lucia Range in areas where vegetation shows relationship to the Broken Hill flora. In addition, the mean monthly range of temperature is 13-14°F (7-7.8°C) greater in the interior. That Pliocene winter temperatures were more mild than those now in the inner valleys of the Santa Lucia Mountains is implied by the abundance of *Persea* in the fossil flora. Furthermore, the polar ice sheet was just commencing to spread, so winters were still mild at this latitude, as *Persea* and *Magnolia* indicate. Low, discontinuous mountains at the site of the present Coast Range barrier would result also in a greater maritime influence over the region, a condition enhanced by the warm, shallow San Joaquin Sea itself. Hence, as depicted in Figure 17, a lower range of temperature in the Pliocene would result in a lower incidence of frost, and in the nearly frost-free climate which is inferred for the flora.

This interpretation is consistent with evidence reviewed by Woodring, Stewart, and Richards (1940) that the marine faunas of the Etchegoin and San Joaquin formations have taxa that indicate warmer seas than at present. The overlying *Pecten coalingensis* zone has a number of taxa (Woodring et al., p. 101) that occur south of Pt. Conception, and some are confined to the warmer waters of the Gulf of California. In addition,

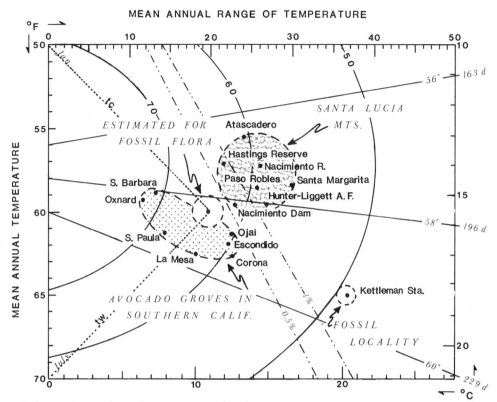

FIG. 16. Comparison of thermal conditions in Kettleman Hills today with those in the inner Santa Lucia Mountains, and in coastal areas where *Persea* is cultivated under nearly frost-free conditions. The thermal field inferred for the Broken Hill flora is based on these data.

mollusks in these formations have their descendants in Japan. Although Woodring et al. (p. 102) state that the possible environmental significance of their survival in Japanese waters is uncertain, the distributional data assembled by Hall (1964), when related to the concept of warmth or effective temperature (Bailey, 1960), seem to clarify the problem. There are fully a dozen Japonic molluscan taxa in the Kettleman Hills fauna that now occur in the Choshian subprovince, in waters that are as warm as those in the California province today. Both have water temperatures of approximately 59.0-64.4°F (15-18°C) for almost four continuous months, and water is not colder than 50°F (10°C). These taxa also reach southward into the outer tropical Kuyushan molluscan subprovince, as well as northward into the cool, temperate Hokkaidan subprovince. However, attainment of the latter distribution was a recent event, representing Quaternary adaptation to colder waters as implied by spreading glaciation and by isotopic evidence concerning the trend of marine paleoclimate (Savin, Douglas, and Stehli, 1975). The increasing ranges of tolerance of taxa with respect to colder water can be viewed in terms of the invasion of new subzones of an expanding marine province. This finds analogy with the adaptive subzones illustrated by Simpson (1953, figs. 49B, 50), and depicted here in Figure 17 for this particular problem.

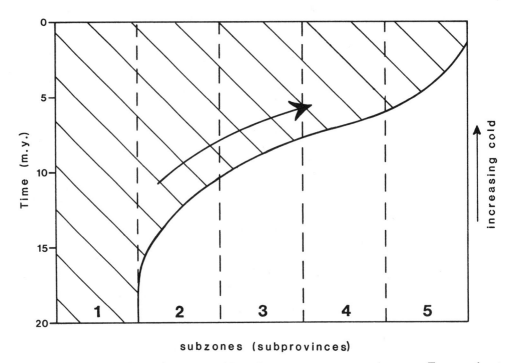

FIG. 17. Spread of taxa into new, colder subzones as temperature decreases. Taxa need not change morphologically to enter a new way of life.

The expanding ranges of tolerance of species for colder water point up the fact that Pliocene (or Miocene) taxa *were not* the same as living species, even though they appear to resemble them morphologically. This is one of the reasons that pre-Quaternary fossil plants have been given fossil names, even though they look like the modern taxa insofar as their leaves, fruits, seeds, or other structures are concerned. They are not the same functionally, and therefore deserve recognition as different taxa. For example, the Miocene and Pliocene live-oak *Quercus wislizenoides* Axelrod cannot be distinguished from leaves of the living *Q. wislizenii* De Candolle. The Neogene species lived under a regime of regular summer rainfall, whereas the descendant *Q. wislizenii* became adapted to a climate of dry summers following the Pliocene. To use the present requirements of *Q. wislizenii* as an indicator of Tertiary climate is clearly misleading. This is also true for other taxa now confined to the California floristic province, whether *Lyonothamnus floribundus*, *Pinus radiata*, or *Sequoiadendron giganteum*, all of which are bradyteles that have also adapted to the new Mediterranean-type climate.

The mollusks with Japanese affinities that earlier were in California waters probably were eliminated here as colder water spread down the eastern side of the Pacific during the Quaternary ice ages, but survived in Japan because that area was bathed by warmer water from southern latitudes. As colder water spread down the northeastern Pacific, effective summer rainfall gradually disappeared from the California region. As a result, trees that are now either in the eastern United States or eastern Asia, or in the southwestern United States and northern Mexico, were eliminated from California, and essentially modern communities remained. Clearly, marine paleotemperature has con-

TABLE 16

Estimated Climatic Conditions under Which
the Broken Hill Flora Lived

Precipitation			
Rainfall			
Total		21-25in.	53-63cm
Winter		18-20in.	45-50cm
Summer		3-5in.	8-12cm
Snowfall		0	0

Temperature			
	Mean annual	60.0°F	15.6°C
*	Mean annual range	20.0°	11.1°
	Mean monthly (see Fig. 1)		
	January	50.0°	10.0°
	February and December	51.3°	10.7°
	March and November	55.0°	12.8°
	April and October	60.0°	15.6°
	May and September	65.0°	18.3°
	June and August	68.7°	20.4°
	July	70.0°	21.1°
	Warmth (W)	58.4°	14.7°
	Days warmer than W	202	202
	Equability (Temperateness)	M 64	M 64
†	Frost frequency	0.3%	0.3%

*Difference between mean temperature of the warmest and coldest months.
†Percentage of hours of the year subject to freezing.

trolled life on land as well as in the sea. The available evidence regarding the nature of the climate under which the Broken Hill flora probably lived is presented in Table 16.

The estimates of temperature are comparable to those based on the ecologic and geochemical analyses of the marine faunas in the Kettleman Hills (Stanton and Dodd, 1970). Their results indicate a marine climate like that now just south of Pt. Conception, with a mean winter sea surface temperature of about 55.5-57°F (13-14°C) and a range of about 5°F (3.7°C), as based on the strontium concentration in the outer layer of *Mytilis* shells in faunal zones below (upper *Pseudocardium* zone) and above (*Acila* zone) the flora. By comparison, land climate along the shore south of Pt. Conception can be judged from that at Santa Barbara airport. The mean temperature there is 58.8°F (14.9°C), the range is 13.7°F (7.6°C), the mean winter temperature for the coldest three months is 52.6°F (11.4°C), and the mean for the coldest month (January) is 51.4°F (10.8°C). Considering the divergent approaches, the natural differences between marine and land climates, and also that the Broken Hill flora is not like that now at Santa Barbara, the results seem in reasonable agreement.

AGE

The Pliocene Age of the Broken Hill flora from the basal part of the San Joaquin Formation is established by its association with rich marine molluscan faunas (Woodring, Stewart, and Richards, 1940). In addition, a few diagnostic mammalian remains in the formation, chiefly from near the middle, indicate that part of the section is Early Blancan. This is shown by *Pleshippus* teeth and *Castor californicus* in the *Pecten* zone, and by a mastodont (*Pliomastodon vexillaris*) just below the *Pecten* zone. All these records were reviewed by Woodring and his associates (1940), and Donald Savage (verbal communication, January 1978) states that no significant new finds have come to light in these strata. Early Blancan mammals have an age of from 4.0 to about 3.0 m.y. (Evernden et al., 1964, p. 164). Inasmuch as the flora occurs below the mammal sites, it is older than 4.0 m.y. A closer assignment is indicated by a radiometric date (4.6 m.y. on zircon, 4.5 m.y. on plagioclase) of a tuff in the upper part of the lower member of the San Joaquin Formation, exposed in Arroyo Dobelgada (sec. 11, T. 22 S, R. 18 E) (Obradovich, Naeser, and Izett, 1978). Since the flora lies several hundred feet below the tuff, it is about 5.0 m.y., or late Hemphillian in age.

LOWLAND VEGETATION
OF THE CENTRAL GREAT VALLEY

Comparison of the Broken Hill flora with the contemporaneous Turlock Lake flora provides a basis for a tentative reconstruction of vegetation at or close to sea level in the central part of the Great Valley of California during the time from about 4.5 to 5.0 m.y. ago. Turlock Lake is on the east side of the Great Valley, in the lowest foothills of the Sierra Nevada 110 miles (117 km) north-northwest of Broken Hill. It was a site of lacustrine deposition that was not more than 80-100 ft. (25-30 m) above sea level. The composition of the two floras is compared in Table 17, which reveals the similarities and differences more clearly than would grouping of all species into a single column, with x's indicating the taxa in common. The following points are especially noteworthy.

1. As for the general vegetation of the region, oak woodland-savanna (or grass) covered the interfluves, and riparian-border forests had a floodplain and lake-border occurrence in each area. To judge from their general similarity and the distribution of the taxa, they extended across the Great Valley north of the San Joaquin embayment, linking both areas.

2. Noteworthy at Turlock Lake is the presence of *Cyperus*, *Juncus*, and *Typha*, which regularly inhabit quiet water along the margins of lakes and sluggish rivers, taxa that are not recorded at Broken Hill. This is understandable from the contrasting environments of deposition—lacustrine or marginal estuarine at Turlock Lake, and floodplain at Broken Hill.

3. Similarities in the floodplain vegetation include the dominant species in each area, notably *Populus garberii* and *Platanus paucidentata*. Four willows occur in these floras, and 3 are in common. The differences in the riparian-border vegetation appear to reflect their settings. Whereas *Persea* is dominant at Broken Hill, it is rare (only 2 specimens) at Turlock Lake. This could be explained by its infrequence near the site of deposition, but it probably reflects other factors. Of these, the substrate with a higher water table and poor drainage, and also a cooler climate as judged from *Arbutus*, *Pinus*, and *Smilax*, seem likely factors accounting for its rarity at Turlock Lake. *Persea* is common at Broken Hill and also at other sites in the San Joaquin and Etchegoin formations wherever plants can be recovered. It occurs chiefly in strata that are coarse and would

have good drainage; and proximity to the sea, in the lee of the low Coast Ranges, would have provided a warmer climate as well.

Only one riparian-border species at Turlock Lake (*Forestiera buchananensis*), was not encountered in the Broken Hill flora. By contrast, the Broken Hill assemblage has 6 riparian-border taxa (*Alnus corrallina, Celtis kansana, Populus alexanderii, P. coalingensis, Salix wildcatensis, Sapindus oklahomensis*) that are not now known from the Turlock Lake flora. The absence of *Celtis, Populus coalingensis* and *Sapindus* at Turlock Lake may be explained by the somewhat lower temperatures than at Broken Hill. In this regard, both *Celtis* and *Sapindus* are in the slightly older and warmer Oakdale flora which lived a short distance north of Turlock Lake (Axelrod, 1944). The richer development of riparian-border species at Broken Hill probably reflects, at least in part, the nature of the substrate. A well-drained, moist sand and gravel made up the floodplain at Broken Hill, whereas at Turlock Lake a poorly drained substrate of claystone and mudstone typified that lacustrine setting. In addition, the proximity of the sea to Broken Hill probably also gave it a somewhat warmer climate, one more favorable for the abundance of *Persea* and other taxa that now occur in warmer regions.

4. As for the woodland, 3 of the oaks (*Quercus douglasoides, Q. pliopalmerii,* and *Q. wislizenoides*) are common to both areas. The other two occur chiefly in drier sites, but the living equivalent of *Q. wislizenoides* which dominates at both sites is a regular member of woodland-grass vegetation on the lower Sierran slopes, and also in the Coast Ranges. As judged from their modern analogues, the oaks at Turlock Lake (*Q. dispersa, Q. douglasoides*) evidently represent taxa of drier requirements than *Q. remingtonii* at Broken Hill. This is understandable because much of the Turlock Lake flora was transported to the area, presumably from sites on the well-drained andesitic sands and gravels of the Mehrten Formation which are exposed nearby. This agrees with the poorer representation of shrubs at Broken Hill, as compared with the Turlock Lake flora which has *Q. dispersa, Q. pliopalmerii, Ceanothus* (2 species), and *Rhamnus* (2 species). They presumably were transported also into the lake from drier, well-drained sites to the east, where the water-table was not as high as on the semihydric delta floodplain at Broken Hill.

5. At Turlock Lake the records of *Amorpha, Arbutus, Pinus* (seeds), *Smilax,* and *Toxicodendron* imply a more mesic woodland and suggest a cooler climate, one not far from the transition to mixed evergreen forest (*Arbutus, Lithocarpus, Quercus chrysolepis*) which lived upslope to the east. The mesic woodland taxa in the Turlock Lake flora probably reached down well-watered valleys toward the site on north exposures, intermingling with the sclerophyllous oaks, as they do today. The only species that suggest this general vegetation zone at Broken Hill are *Lithocarpus* and *Q. remingtonii.* Their presence there implies that mixed evergreen forest—which had dominated the lowlands of this area in the Miocene (Renney, 1969)—probably clothed the Coast Ranges directly west. The evidence for a somewhat moister climate at Turlock Lake seems understandable on the basis of its position on the lower windward slopes of the Sierra Nevada fully 100 miles (106 km) north, where moister, cooler conditions might be expected much as they occur there today. By contrast, the Broken Hill flora was situated farther south in the lee of the low, ancestral Coast Ranges, and would naturally be somewhat drier and also warmer. A warmer climate there would result not only in a more nearly frost-free climate, but in more effective summer rainfall and in the occur-

TABLE 17

Comparison of Represented Taxa in Vegetation Zones
of the Broken Hill and Turlock Lake floras

Turlock Lake Flora	*Broken Hill Flora*

Lacustrine

Cyperaceae
    Cyperus sp.

Juncaceae
    Juncus sp.

Typhaceae
    Typha lesquereuxii

Riparian-Border

Liliaceae
    Smilax remingtonii

Salicaceae   Salicaceae
                                                          Populus alexanderii
                                                          Populus coalingensis
    Populus garberii......................*Populus garberii*
    Salix edenensis ....*Salix edenensis*
    Salix hesperia..........................*Salix hesperia*

    Salix laevigatoides ....*Salix laevigatoides*
                                                              Salix wildcatensis

                                                         Betulaceae
                                                            Alnus corrallina

                                                              Ulmaceae
                                                            Celtis kansana
                                                            Ulmus affinis

                                                              Magnoliaceae
                                                            Magnolia corrallina

Lauraceae
    Persea coalingensis...................*Persea coalingensis*
    Umbellularia salicifolia..............*Umbellularia salicifolia*

Platanaceae   Platanaceae
    Platanus paucidentata ....*Platanus paucidentata*

                                                              Sapindaceae
                                                            Sapindus oklahomensis

Oleaceae
    Forestiera buchananensis

Woodland-Savanna (Grass)

Fagaceae   Fagaceae
    Quercus dispersa
    Quercus douglasoides ....*Quercus douglasoides*
                                                              Quercus lakevillensis
    Quercus pliopalmerii ....*Quercus pliopalmerii*

Quercus wislizenoides ....*Quercus wislizenoides*

Fabaceae　　　　　　　　　　　Fabaceae
　Amorpha condonii
　　　　　　　　　　　　　　　　　Amorpha oklahomensis

Anacardiaceae
　Toxicodendron franciscana

Rhamnaceae
　Ceanothus tuolumnensis
　Ceanothus turlockensis
　Rhamnus moragensis
　Rhamnus precalifornica

Mixed Evergreen Forest (Marginal)

Pinaceae
　Pinus sturgisii

Liliaceae
　Smilax remingtonii

　　　　　　　　　　　Fagaceae
　　　　　　　　　　　　Lithocarpus klamathensis
　　　　　　　　　　　　Quercus remingtonii

Ericaceae
　Arbutus matthesii

---

rence of more numerous summer-rain indicators, as shown by the records there of *Persea* (dominant), *Magnolia, Sapindus* and others.

In summary, comparison of the Broken Hill and Turlock Lake floras reveals that oak woodland-savanna (grass) and riparian-border vegetation covered the central Great Valley in areas north of the marine embayment during the transition from Hemphillian into Blancan time (4.5 to 3.5 m.y.). These vegetation zones were richer in taxa than are the surviving communities in California, chiefly because there was still sufficient summer rainfall for taxa that are no longer here. While a number of species were common to both areas, there were important differences in composition. These appear to reflect the differences in the paleogeographic settings, edaphic conditions, and the local climates under which they lived.

SUMMARY

The Broken Hill flora from the San Joaquin Formation in the Kettleman Hills includes 22 species, all similar to living ones. The taxa contributed to vegetation of a delta floodplain that appeared briefly in the San Joaquin Sea during deposition of the Cascajo Member. At that time the Coast Ranges had not yet been elevated appreciably, and the seaway trended northwest to the Pacific, across the trend of the present Coast Ranges.

The ecologic occurrence of modern species similar to the fossils is paralleled by vegetation in the inner Santa Lucia Mountains, where species of oak woodland-savanna and riparian-border vegetation intermingle on the floodplain. Comparative climatic data indicate that rainfall was near 20-23 in. (50-58 cm), or fully 14-17 in. (35-43 cm) more than that at the site today. Several taxa in the flora, including the genera *Persea* and *Sapindus*, point to effective summer rain, probably totalling at least 1 in. (3 cm) per summer month, followed immediately by heavy fall rains. It is estimated that temperatures were more like those now in the Coast Ranges to the west, with a mean annual temperature near 60°F (15.6°C) and an annual range near 20°F (11°C). Uplift of a continuous and higher Coast Range barrier following deposition of the overlying Tulare Formation gave the Kettleman Hills area its interior position and brought to it a drier, hotter, less equable climate. Mean temperature has increased fully 5°F (2.8°C), and the range of temperature is now about 15°F (8.3°C) greater. The age of the flora is established as Pliocene (transitional Hemphillian-Blancan), or 4.5 m.y. old.

SYSTEMATIC DESCRIPTIONS

Family SALICACEAE
Populus alexanderii Dorf
(Plate 17, fig. 1)

Populus alexanderii Dorf, Carnegie Inst. Wash. Pub. 412, p. 75, pl. 6, figs. 10, 11; pl. 7, fig. 3, 1930; Axelrod, Univ. Calif. Pub. Geol. Sci., vol. 34, p. 128, pl. 19, figs. 1-11, 1958.

Two ovate to oval leaf impressions with a finely toothed margin have the typical venation of this species, which resembles the leaves of *P. trichocarpa* in the coastal region from central California near Monterey Bay southward into southern California. The leaves differ significantly from the taxon identified as *P. trichocarpa* in the mountains of California and northward, a taxon that seems more nearly allied to *P. balsamifera* and *P. hastata*. The taxon from the coastal strip of central and southern California lives under a mild winter climate, and one of only moderate rainfall.

Collection: U.C. Mus. Pal., Paleobot. Ser., Broken Hill, hypotype no. 6138.

Populus coalingensis n. sp.
(Plate 16, fig. 1)

Description. Leaf blade cuneiform, cuneate below and bitten above; blade 3.2 cm long and 4.0 cm broad; petiole slender, 3.0+ cm long. Secondaries diverging at relatively low-medium angles, repeatedly bifurcating upward to supply the distal margin, the basal pair scarcely as strong as primaries; tertiaries poorly preserved, apparently irregularly cross-percurrent; marginal teeth on outer edge of blade relatively small; texture evidently medium.

Discussion. This species is based on a single specimen that seems closely allied to the living *P. euphratica* Oliver, which ranges from central and western Asia into Asia Minor and adjacent Africa. The only other American fossil species that shows relationship to *P. coalingensis* is *P. gallowayii* MacGinitie, from the Kilgore flora of northwestern Kansas (MacGinitie, 1962). It differs from *P. coalingensis* and *P. euphratica* in having much larger and fewer teeth, and apparently does not include all the variation of leaves of *euphratica*, which may simulate those of *Salix*. These American records of taxa that seem allied to *euphratica* are of high interest, for they provide another link to the flora of the interior of Asia, represented in this case by members of a very unique section of the genus *Populus*.

Collection: U.C. Mus. Pal., Paleobot. Ser., Broken Hill, holotype no. 6139.

Populus garberii Axelrod
(Plate 16, figs. 3-5)

Populus garberii Axelrod, Univ. Calif. Pub. Geol. Sci. 121, p. 126, pl. 10, figs. 1-3, 1980 (see synonymy).

Numerous leaves of this species, which are similar to those in the Turlock Lake flora, are represented in the collection from Broken Hill. They are similar to the leaves of *P. tremula* of Eurasia, differing from those of *P. grandidentata* chiefly in their more numerous and smaller marginal teeth. Its occurrence in these and in other floras in central California, notably the Neroly and Black Hawk Ranch, seems to have been chiefly on the floodplains.

Collection: U.C. Mus. Pal., Paleobot. Ser., Broken Hill, hypotype nos. 6140-6142; homeotype nos. 6143-6156.

Salix edenensis Axelrod

Salix edenensis Axelrod, Carnegie Inst. Wash. Pub. 590, p. 101, 1950.
Salix sp. Dorf. Axelrod, *ibid.* 476, p. 171, pl. 4, fig.7, 1937.
Salix payettensis Axelrod, *ibid.* 553, p. 254, pl. 43, fig. 9; Axelrod, *ibid.* 590, p. 203, pl. 4, figs. 3 and 8, 1950.

All of the above-listed material represents a willow of the sandbar willow alliance, including species like *S. exigua* Nuttall, *S. interior* Rowlee, and *S. fluvatilis* Nuttall. Inasmuch as the living taxa are distinguished by characters that are not preserved as fossil, and since the leaves of these taxa are all quite similar, it seems best to group all of the fossil species into a generalized willow of the sandbar alliance.

The slender leaves in the Broken Hill flora that represent this species are similar to those of *S. exigua* and *S. fluvatilis*, and like them imply a floodplain habitat.

Collection: U.C. Mus. Pal., Paleobot. Ser., Broken Hill, homeotype nos. 6157, 6158.

Salix hesperia (Knowlton) Condit
(Plate 17, fig. 8)

Salix hesperia (Knowlton) Condit, Carnegie Inst. Wash. Pub. 553, p. 41, pl. 4, fig. 7, 1944 (see synonymy).

Several large lanceolate leaves in the flora, mostly incomplete, represent this species, which is allied to the living *S. lasiandra* Bentham. As in the living species, the fossils display numerous camptodrome secondaries and intersecondaries, the tip is acuminate, the base acute to rounded, the petiole long and of moderate thickness, and the margin varies from serrate to entire. *S. lasiandra* is one of the more widely distributed willows in the Pacific region west of the Sierra-Cascade axis, extending from southern British Columbia into southern California, but with outliers in eastern Washington and Idaho in the Columbia embayment, and also in central New Mexico and in Colorado. The variety *caudata* (Nuttall) Sudworth, which replaces it over most of the intermontane region, has smaller leaves.

Collection: U.C. Mus. Pal., Paleobot. Ser., Broken Hill, hypotype no. 6159; homeotype nos. 6160, 6161.

Salix laevigatoides Axelrod
(Plate 17, figs. 2-5)

Salix laevigatoides Axelrod, Carnegie Inst. Wash. Pub. 590, p. 55, pl. 2, fig. 10, 1950.

Numerous leaf impressions in the flora are similar to those produced by the living *S. laevigata* Bebb. This common willow is widely distributed in western California from the

Oregon border southward, and also in the Great Valley. It reaches into western Arizona, and also up the Colorado River drainage into southern Nevada and adjacent Utah. In the fossils, the variation in the angle of divergence of the secondaries is well matched by that of the living species, as is the nature of the margin, which ranges from essentially entire to finely serrate.

Collection: U.C. Mus. Pal., Paleobot. Ser., Broken Hill, hypotype nos. 6162-6165; homeotype nos. 6166-6172; 6262-6265.

Salix wildcatensis Axelrod
(Plate 17, figs. 9 and 10)

Salix wildcatensis Axelrod, Carnegie Inst. Wash. Pub. 553, p. 132, 1944 (see synonymy and discussion); Chaney, *ibid.*, p. 341, pl. 58, fig. 2; pl. 59, figs. 1-3 (not fig. 4, which is *Castanopsis sonomensis* Axelrod), 1944.

The typical leaves of this species are represented in the flora by impressions that are long lanceolate, entire, and with blunt to acute tips and acute bases. They are similar in all respects to leaves of the common and widespread *S. lasiolepis* Bentham, which ranges from California into Arizona and northern Mexico. It inhabits diverse vegetation belts, ranging from the redwood and mixed conifer forests into woodland vegetation, and to the margins of the desert.

Collection: U.C. Mus. Pal., Paleobot. Ser., Broken Hill, hypotype nos. 6173 and 6174; homeotype nos. 6175-6177.

Family BETULACEAE
Alnus corrallina Lesquereux
(Plate 17, fig. 11)

Alnus corrallina Lesquereux, U.S. Geol. Surv. Terr. Rept. vol. 8, p. 243, pl. 51, figs. 1-3, 1883.

A single, well-preserved leaf in the flora, rhombic in outline, with numerous secondaries, and a strong cross-percurrent network of tertiary veins, seems assignable to this species. Among living alders, *A. rhombifolia* Nuttall has leaves similar to the fossil. This is a widely distributed species in the Pacific states, reaching inland up the Columbia River basin into the panhandle of northern Idaho.

This species seems to differ from *A. merriamii* Dorf from the Calabasas locality (Dorf, 1930, pl. 8, fig. 7), which is fragmentary. Whether it is different from the specimen from the Wildcat Formation (Dorf, 1930, pl. 8, fig. 6) is uncertain and cannot be determined definitely until a larger suite of *A. merriamii* is recovered. If they prove to be similar, *merriamii* becomes a synonym of *corrallina*.

Collection: U.C. Mus. Pal., Paleobot. Ser., Broken Hill, hypotype no. 6178.

Family FAGACEAE
Lithocarpus klamathensis (MacGinitie) Axelrod

Lithocarpus klamathensis (MacGinitie) Axelrod, Carnegie Inst. Wash. Pub. 553, p. 197, pl. 37, fig. 4 (see synonymy).

Quercus simulata Knowlton. Condit, Carnegie Inst. Wash. Pub. 553, p. 45, pl. 3, fig. 3; Axelrod, Univ. Calif. Pub. Geol. Sci. 33, p. 291, pl. 13, fig. 12; pl. 27, figs. 1-4; Axelrod, Univ. Calif. Pub. Geol. Sci. 39, p. 223, pl. 48, fig. 3, 1962.

This species is represented by the distal half of a thick, lanceolate leaf 4 cm long and 1.5 cm wide. The heavy teeth are widely spaced, some blunt but others sharp. Similar leaves are produced by *L. densiflorus* Hooker and Arnott in the drier, interior parts of California.

Specimens previously referred to *Quercus simulata* in several California and Nevada Neogene floras are transferred here to *Lithocarpus*; they show no relation to *Quercus myrsinaefolia* Blume, as previously implied by others.

Collection: U.C. Mus. Pal., Paleobot. Ser., Broken Hill, homeotype no. 6179.

Quercus douglasoides Axelrod
(Plate 17, figs. 6 and 7; Plate 18, fig. 1)

Quercus douglasoides Axelrod, Carnegie Inst. Wash. Pub. 553, p. 198, pl. 37, figs. 7-10, 1944 (see synynomy); Axelrod, *ibid.* 590, p. 205, pl. 5, fig. 4, 1950.

Three leaves in the flora are referred to this species, which is characterized by broadly oval to ovate leaves that are shallowly lobed, and the lobes may be either with or without a small prickle. Among living species, the fossils are similar to leaves of the blue oak, *Q. douglasii* Hooker and Arnott. This hardy tree, deciduous in the cold season, is distributed around the Great Valley of California in the Sierran foothills and in the drier Coast Ranges, reaching southward with digger pine to the margins of southern California, as on Liebre Mountain south of Sandberg and in Santa Ynez Valley north of Santa Barbara.

Collection: U.C. Mus. Pal., Paleobot. Ser., Broken Hill, hypotype nos. 6180-6182.

Quercus lakevillensis Dorf
(Plate 18, figs. 2 and 3)

Quercus lakevillensis Dorf, Carnegie Inst. Wash. Pub. 412, p. 82, pl. 8, figs. 4, 5, 1930; Axelrod, *ibid.* 590, p. 58, pl. 3, fig. 4, 1950.

Two ovate leaf impressions referred to this species have 3-4 secondaries that are distinctly wavering in their outward course, the marginal sinuses are broadly scalloped, and the teeth are not as prominently developed as in *Q. wislizenoides* or *Q. pliopalmerii*. Among living species, *Q. agrifolia* has leaves similar to those of the fossil. It is widely distributed in the outer Coast Ranges from near the Sonoma-Mendocino Counties line southward into northern Baja California. It has a scattered occurrence in the higher parts of the inner Coast Ranges, with relict stations in the Diablo and Temblor ranges north and south of the Kettleman Hills.

Collection: U.C. Mus. Pal., Paleobot. Ser., hypotype nos. 6266 and 6183.

Quercus pliopalmerii Axelrod
(Plate 18, figs. 8-10)

Quercus pliopalmerii Axelrod, Carnegie Inst. Wash. Pub. 476, p. 174, pl. 5, figs. 1-3; Axelrod, *ibid.* 590, p. 147, pl. 2, figs. 6 and 9, 1950.

The leaf impressions of this species are readily separated from those of *Q. wislizenoides* by having only 3-4 rather than 7-9 secondaries; by their higher angle of divergence; by the strongly crisped nature of the leaves; and by the broad, shallow sinuses between the sharply pointed, long teeth. The modern species with leaves most similar to it is *Q. dunnii* (= *palmerii*) of interior southern California and in southern Arizona, which also has relict outposts in the inner south Coast Ranges.

Collection: U.C. Mus. Pal., Paleobot. Ser., Broken Hill, hypotype nos. 6184-6186; homeotype no. 6187.

Quercus remingtonii Condit
(Plate 19, figs. 1-4)

Quercus remingtonii Condit, Carnegie Inst. Wash. Pub. 553, p. 44, pl. 6, figs. 1-3, 5-7; pl. 7, figs. 1, 2; pl. 8, fig. 2, 1944.

The leaves of this species are common in the flora, and are highly variable in shape, ranging from obovate to lanceolate or oblanceolate. The lobes vary from simple to complex, and are regularly surmounted by bristle-tipped teeth. They fall readily within the variation displayed by the living *Q. morheus* Kellogg, which is a hybrid of *Q. kelloggii* Newberry and *Q. wislizenii* A. De Candolle. Although *Q. wislizenoides* is represented in the flora by relatively abundant leaves, those of the other parent have not been found. The tree may well have lived a short distance to the west, in the foothills of the ancestral Coast Ranges where climate was a little cooler and more mesic. Even in its present distribution, *Q. morheus* may frequently be found well removed from one of its parents.

Collection: U.C. Mus. Pal., Paleobot. Ser., Broken Hill, hypotype nos. 6188-6191; homeotype nos. 6192-6203.

Quercus wislizenoides Axelrod
(Plate 18, figs. 4-7)

Quercus wislizenoides Axelrod, Carnegie Inst. Wash. Pub. 553, p. 136, pl. 29, figs. 4-9, 1944; Axelrod, *ibid.* 162, pl. 33, figs. 2, 5, 1944. Axelrod, *ibid.* 590, p. 59, pl. 2, figs. 11-12, 1950; Axelrod, *ibid.* 590, p. 148, pl. 3, figs. 1-3, 1950; Axelrod, *ibid.* 590, p. 205, pl. 5, fig. 5, 1950.

The leaves of this oak are among the common specimens in the Broken Hill flora. They show a degree of variability that is closely matched by leaves of the living *Q. wislizenii* of California. It is distributed around the Great Valley of California in the lower foothills of the Sierra Nevada and in the inner Coast Ranges, reaching south through the Transverse and Peninsular ranges into the San Pedro Martir Mountains of Baja California. It occurs also in the outer Coast Ranges of California—but that form, which is a regular member of the mixed evergreen forest, is quite different and seems worthy of subspecific recognition. The fossils from Broken Hill and other floras in California are most nearly allied to the taxon in the interior parts of its distribution, where it is often a dominant member of the woodland vegetation. Current studies suggest that the leaves recorded from the Late Miocene floras of west-central Nevada (Axelrod, 1956), which are regularly larger than the California specimens, are more nearly allied to the variety now in the mixed evergreen forest.

Collection: U.C. Mus. Pal., Paleobot. Ser., Broken Hill, hypotype nos. 6204-6207; homeotype nos. 6208-6220.

Family ULMACEAE
Celtis kansana Chaney and Elias
(Plate 16, fig. 2)

Celtis kansana Chaney and Elias, Carnegie Inst. Wash. Pub. 476, p. 38, pl. 5, figs. 1-5, 1936; Condit, *ibid.* 553, p. 77, pl. 14, fig. 4, 1944; Axelrod, *ibid.* 553, p. 136, pl. 30, fig. 1, 1944.

The basal third of a leaf with asymmetrical primaries seems referable to this species, which is similar to *C. reticulata*. This is a shrub of the southwestern United States that has a relict occurrence in San Diego and Riverside counties, and also along the Kern River near Caliente, where it is associated with desert-border species. An undescribed specimen of this species is also in the Kreyenhagen Hills flora to the west (Dorf, 1930), which occurs in the *Pecten* zone a few hundred feet stratigraphically above the Broken Hill flora.

Collection: U.C. Mus. Pal., Paleobot. Ser., Broken Hill, hypotype no. 6221; also homeotype no. 334 from Loc. 164 in Kreyenhagen Hills.

Ulmus affinis Lesquereux
(Plate 18, fig. 11)

Ulmus affinis Lesquereux, Harvard Mus. Comp. Zool. Mem. vol. 6, p. 16, pl. 4, fig. 4 (part), 1878; Tanai and Wolfe, U.S. Geol. Surv. Prof. Paper 1026, p. 4, pl. 3, figs. B, D, E, G, 1977.

As revised recently by Tanai and Wolfe (1977), species of *Ulmus* formerly referred to *U. californica* Lesquereux are now placed under *U. affinis* Lesquereux. The specimen in the Broken Hill flora is not complete, but its major features seem sufficiently preserved to indicate that it represents a species with leaves similar to those of the living *U. americana* of the eastern United States. In western parts of its distribution, it reaches well out into the plains along river valleys into areas of subhumid climate, and in proximity to sclerophyll woodland, as in the Edwards Plateau.

Collection: U.C. Mus. Pal., Paleobot. Ser., Broken Hill, hypotype no. 6222.

Family MAGNOLIACEAE
Magnolia corrallina Chaney and Axelrod
(Plate 21, figs. 5 and 6)

Laurus princeps Heer. Lesquereux, Rept. U.S. Geol. Surv. Terr., vol. 8, p. 250, pl. 58, fig. 2, 1883.
Magnolia corrallina Chaney and Axelrod, Carnegie Inst. Wash. Pub. 617, p. 175 (see discussion), 1959.

Two leaf impressions in the flora are referred to this species, which resembles leaves of *M. virginiana* Linnaeus as well as the slender leaves of *M. grandiflora* Linnaeus. Its magnolian affinity is evident in the numerous camptodrome secondaries that diverge at medium-high angles, the numerous subsecondaries that disappear laterally, and in the even quaternary mesh made by the finer nervation. The largest leaf is 12.5 cm long (tip missing) and 3.5 cm broad, is long-elliptic in outline, and has a medium petiole of which 1.5 cm is preserved. The living taxa are members of the eastern mixed forests, especially along the Atlantic coastal plain, where winters are moderated and deep rich soils are present in the bottomlands.

The occurrence of *Magnolia* in this area in the Pliocene is not surprising, in view of the abundance of *Persea* in the same beds. Further, it is represented in the Late Miocene Neroly flora from Corrall Hollow, situated about 135 miles (215 km) northwest. In addition, a similar species is in the Early Pleistocene Soboba flora of southern California. These are clearly relicts of the Tertiary, and indicate the persistence of significant (though small) amounts of summer rainfall at this late date.

Collection: U.C. Mus. Pal., Paleobot. Ser., Broken Hill, hypotype nos. 6223 and 6224.

Family LAURACEAE
Persea coalingensis (Dorf Axelrod)
(Plate 19, fig. 7; pl. 20, figs. 2-4)

Persea coalingensis (Dorf) Axelrod, Carnegie Inst. Wash. Pub. 553, p. 132, 1944 (see synonymy); Axelrod, *ibid.* 590, p. 61, pl. 3, figs. 5, 6, 1950; Axelrod, *ibid.* 590, p. 148, pl. 3, fig. 9, 1950; Axelrod, *ibid.* 590, p. 206, pl. 5, fig. 9, 1950.

The large elliptic leaves of avocado dominate the flora from Broken Hill and are also represented in collections at several sites in the underlying Etchegoin Formation, as well as in the Jacalitos Formation in the Kreyenhagen Hills directly west. The chief uncertainty regarding these specimens, and others in the Pliocene of California, is whether or not more than one species may be represented. Some are broadly elliptic whereas others are narrowly so, yet there appear to be intergradations. Until larger collections can be made, it seems unwise to recognize two taxa. In any event, such a procedure would scarcely alter the paleoecological relations indicated for the flora. Among modern species, the leaves produced by *P. podadenia* Blake of northern Mexico are similar to the fossils, though resemblance to *P. borbonia* and *P. pubescens* of the southeastern United States is also evident.

Collection: U.C. Mus. Pal., Paleobot. Ser., Broken Hill, hypotype nos. 6225-6228; homeotype nos. 6229-6242.

Umbellularia salicifolia (Lesquereux) Axelrod
(Plate 19, figs. 5 and 6)

Umbellularia salicifolia (Lesquereux) Axelrod, Carnegie Inst. Wash. Pub. 512, p. 102, pl. 8, fig. 4 (see synonymy); Axelrod, *ibid.* 590, p. 62, pl. 3, fig. 3, 1950.

The typical narrowly lanceolate leaves of California laurel are present in the collection. They are easily separated from those of *Persea* by their more slender shape; by the secondary venation, which is not so prominently camptodrome because the secondaries are relatively straight in their outward course, rather than arching; and by the tertiary venation, which is much coarser in *Umbellularia*. The species ranges from southwestern Oregon southward through the Coast Ranges, and is scattered in the Transverse and Peninsular ranges into central San Diego County. It is more scattered and less frequent in the Sierra Nevada, chiefly in the woodland belt.

Collection: U.C. Mus. Pal., Paleobot. Ser., Broken Hill, hypotype nos. 6243 and 6244.

Family PLATANACEAE
Platanus paucidentata Dorf
(Plate 20, fig. 1; pl. 21, fig. 7)

Platanus paucidentata Dorf, Carnegie Inst. Wash. Pub. 412, p. 94, pl. 10, figs. 4, 9; pl. 11, fig. 1; pl. 12, fig. 1, 1930; Axelrod, *ibid.* 476, p. 174, pl. 5, figs. 4, 5, 1937; Condit, *ibid.* 553, p. 48, pl. 9, fig. 1; pl. 10, fig. 4, 1944; Axelrod, *ibid.* 590, p. 62, pl. 4, figs. 1, 8, 1950; Axelrod, *ibid.* 590, p. 207, pl. 5, fig. 7; pl. 6, fig. 1, 1950.

Impressions of leaves of this species are common in the flora. However, in view of the large size and spreading lobes of the leaves and the poorly indurated nature of the sandstones in which they occur, it is difficult to collect complete specimens. The fossils are similar to leaves of the living *P. racemosa*, which occurs as isolated stands in the Kreyenhagen Hills, chiefly along the larger stream-courses that drain the Coast Ranges to the west. It is a common member of woodland vegetation from central California southward, reaching up to the lower margins of conifer forest in the southern Sierra Nevada, southern California, and northern Baja California. It mingles with coast redwood forest in the Santa Cruz and Santa Lucia mountains, but does not range farther north in the Coast Ranges, probably because climate is too cool there.

Collection: U.C. Mus. Pal., Paleobot. Ser., Broken Hill, hypotype nos. 6245 and 6246; homeotype nos. 6247-6252.

Family FABACEAE
Amorpha oklahomensis (Berry) Axelrod
(Plate 21, figs. 1-4)

Amorpha oklahomensis (Berry) Axelrod, Univ. Calif. Pub. Geol. Sci., vol. 33, pp. 302-303, 1956 (see discussion).
Diospyros pretexana Chaney and Elias, Carnegie Inst. Wash. Pub. 476, p. 44, pl. 7, figs. 6, 8, 1936.

Salix coalingensis Dorf. Chaney and Elias, *ibid.* 476, pl. 4, figs. 1 and 3 (not fig. 6, which is *Sapindus oklahomensis* Berry), 1936.

Bumelia oklahomensis Berry, Prof. U.S. Nat. Mus., vol. 54, p. 634, pl. 94, fig. 1, 1918.

As discussed earlier (Axelrod, 1956, p. 302), all of the above-listed leaflets in the Beaver flora of Oklahoma represent a legume and, contrary to Brown (1940), are more nearly allied to *Amorpha* than to *Robinia*. They are similar to the leaflets of *A. fruticosa* Linnaeus, a widely distributed shrub ranging from southern California eastward to the Great Plains. Numerous leaflets in the Broken Hill flora are similar to those of *A. fruticosa*, and also to the variation displayed by the species from the Beaver flora. They are therefore referred to that species.

Collection: U.C. Mus. Pal., Paleobot. Ser., Broken Hill, hypotype nos. 6253-6256; homeotype nos. 6257-6260.

Family SAPINDACEAE
Sapindus oklahomensis Berry
(Plate 20, fig. 5)

Sapindus oklahomensis Berry, Prof. U.S. Nat. Mus., vol. 54, p. 632, pl. 95, figs. 1, 2, 1918; Chaney and Elias, Carnegie Inst. Wash. Pub. 476, p. 43, pl. 7, figs. 1-3, 1936; Axelrod, *ibid.* 590, p. 111, pl. 3, fig. 4, 1950 (see synonymy and discussion).

The basal part of an asymmetric leaflet in the flora is referred to this species, which has similar leaflets. The secondaries are subparallel, camptodrome, and diverge at moderate angles. The specimen is of medium texture, the petiolule is short and thick, and its total length was no doubt fully twice that of the preserved material.

The leaflets of the living *S. drummondii* Hooker and Arnott resemble the fossil. Soapberry ranges from Arizona eastward to the Great Plains and southward into Mexico. Near the border, it is replaced by the related *S. saponaria* Linneus, which is American-tropical in distribution. *S. saponaria* differs from *S. drummondii* in having leaflets with blunt to acute tips. Some authorities regard *drummondii* as a variety of *saponaria*, but this is scarcely acceptable in view of the antiquity of the taxon—reaching back to the Eocene.

Collection: U.C. Mus. Pal., Paleobot. Ser., Broken Hill, hypotype no. 6261.

REFERENCES CITED

Axelrod, D.I.
  1944a The Oakdale flora. Carnegie Inst. Wash. Pub. 553: 147-166.
  1944b The Sonoma flora. Carnegie Inst. Wash. Pub. 553: 167-206.
  1950 Studies in Late Tertiary paleobotany. Carnegie Inst. Wash. Pub. 590. 323 pp.
  1956 Mio-Pliocene floras from west-central Nevada. Univ. Calif. Pub. Geol. Sci. 33. 316 pp.
  1966 The Pleistocene Soboba flora of southern California. Univ. Calif. Pub. Geol. Sci. 60. 109 pp.
  1976 History of the conifer forests, California and Nevada. Univ. Calif. Pub. Botany 70. 62 pp.
Bailey, H.P.
  1960 A method of determining the warmth and temperateness of climate. Geografisker Annaler 42: 1-16.
  1964 Toward a unified concept of the temperate climate. Geogr. Rev. 54: 516-545.
Brown, R.W.
  1940 New species and changes of name in some American fossil floras. Wash. Acad. Sci. Jour. 30: 344-356.
Condit, C.
  1944 The Remington Hill flora. Carnegie Inst. Wash. Pub. 553: 21-55.

Dorf, E.
    1930 Pliocene floras of California. Carnegie Inst. Wash. Pub. 412: 1-108.
Evernden, J.F., D.E. Savage, G.H. Curtis, and G.T. James
    1964 Potassium-argon dates and the Cenozoic mammalian chronology of North America. Amer. Jour. Sci. 262: 145-198.
Galehouse, J.S.
    1967 Provenance and paleocurrents of the Paso Robles Formation, California. Geol. Soc. Amer. Bull. 78: 951-978.
Hackel, O.
    1966 Summary of the geology of the Great Valley. Calif. Div. Mines and Geol. Bull. 190: 215-238.
Hall, C.A.
    1964 Shallow-water marine climates and molluscan provinces. Ecology 45: 226-234.
Jepson, W.L.
    1910 The Silva of California. Univ. Calif. Memoir 2: 1-480.
MacGinitie, H.D.
    1962 The Kilgore flora, a Late Miocene flora from northern Nebraska. Univ. Calif. Pub. Geol. Sci. 35: 67-158.
Obradovich, J.D., C.W. Naeser, and G.A. Izett
    1978 Geochronology of Late Neogene marine strata in California. Stanford Univ. Pub. Geol. Sci. 14: 40-41.
Raven, P.H., and D.I. Axelrod
    1978 Origin and relationships of the California flora. Univ. Calif. Pub. Botany 72. 144 pp.
Reed, R.D.
    1933 The Geology of California. Amer. Assoc. Petrol. Geol. Tulsa, Oklahoma. 355 pp.
Renney, K.M.
    1969 The Miocene Temblor flora of west-central California. M.A. thesis, Univ. California, Davis. 83 pp.
Savin, S.M., R.G. Douglas, and F.G. Stehli
    1975 Tertiary marine paleotemperatures. Geol. Soc. Amer. Bull. 86: 1499-1510.
Simpson, G.G.
    1953 The Major Features of Evolution. New York: Columbia Univ. Press. 434 pp.
Stanton, R.J., Jr., and J.R. Dodd
    1970 Paleoecologic techniques—comparison of faunal and geochemical analyses of Pliocene paleoenvironments, Kettleman Hills, California. Jour. Paleo. 44: 1092-1121.
Stewart, R.B.
    1946 Geology of Reef Ridge, Coalinga district, California. U.S. Geol. Surv. Prof. Paper 205-C: 81-115.
Tanai, T., and J.A. Wolfe
    1977 Revision of *Ulmus* and *Zelkova* in the middle and late Tertiary of western North America. U.S. Geol. Surv. Prof. Paper 1026: 1-14.
Woodring, W.P., R. Stewart, and R.W. Richards
    1940 Geology of the Kettleman Hills Oil Field, California. U.S. Geol. Surv. Prof. Paper 195. 170 pp.

Chapter IV Plates

PLATE 15

Fossil locality and modern vegetation similar to the Broken Hill flora

FIG. 1. Looking east at Broken Hill, situated at the south end of North Dome, Kettleman Hills, California. The plant-bearing beds occur in the Cascajo Member of the San Joaquin Formation, which forms the sandstone cliff on the crest of the low ridge. Note that it has been displaced by a small fault that trends northeasterly along the trace of the cattle trail.

FIG. 2. Modern vegetation in the Santa Lucia Mountains is allied to that represented in the fossil flora. This view is to the southwest from the site of the Nacimiento weather station (14-year record). The Nacimiento River floodplain supports species of *Alnus*, *Platanus*, *Populus*, and *Salix*. The bordering plain and low hills are covered with oak-grassland. Mixed evergreen forest and chaparral blanket the slopes. Some 13 taxa in this area have allied species in the fossil flora.

PLATE 16
Broken Hill fossils

FIG. 1. *Populus coalingensis* Axelrod. Holotype no. 6139.
FIG. 2. *Celtis kansana* Chaney and Elias. Hypotype no. 6221.
FIGS. 3-5. *Populus garberii* Axelrod. Hypotype nos. 6140-6142.

PLATE 17
Broken Hill fossils

FIG. 1. *Populus alexanderii* Dorf. Hypotype no. 6138.

FIGS. 2-5. *Salix laevigatoides* Axelrod. Hypotype nos. 6162-6165.

FIGS. 6, 7. *Quercus douglasoides* Axelrod. Hypotype nos. 6182, 6181.

FIG. 8. *Salix hesperia* (Knowlton) Condit. Hypotype no. 6159.

FIGS. 9, 10. *Salix wildcatensis* Axelrod. Hypotype nos. 6174, 6173.

FIG. 11. *Alnus corrallina* Lesquereux. Hypotype no. 6178.

PLATE 18

Broken Hill fossils

FIG. 1. *Quercus douglasoides* Axelrod. Hypotype no. 6180.

FIGS. 2, 3. *Quercus lakevillensis* Dorf. Hypotype nos. 6266, 6183.

FIGS. 4-7. *Quercus wislizenoides* Axelrod. Hypotype nos. 6204-6207.

FIGS. 8-10. *Quercus pliopalmerii* Axelrod. Hypotype nos. 6184-6186.

FIG. 11. *Ulmus affinis* Lesquereux. Hypotype no. 6222.

PLATE 19
Broken Hill fossils

FIGS. 1-4. *Quercus remingtonii* Condit. Hypotype nos. 6188-6191.

FIGS. 5, 6. *Umbellularia salicifolia* (Lesquereux) Axelrod. Hypotype nos. 6243, 6244.

FIG. 7. *Persea coalingensis* (Dorf) Axelrod. Hypotype no. 6228.

PLATE 20
Broken Hill fossils

FIG. 1. *Platanus paucidentata* Dorf. Hypotype no. 6246.

FIGS. 2-4. *Persea coalingensis* (Dorf) Axelrod. Hypotype nos. 6225-6227.

FIG. 5. *Sapindus oklahomensis* Berry. Hypotype no. 6261.

PLATE 21
Broken Hill fossils

FIGS. 1-4. *Amorpha oklahomensis* (Berry) Axelrod. Hypotype nos. 6253-6256.

FIGS. 5, 6. *Magnolia corrallina* Chaney and Axelrod. Hypotype nos. 6224, 6223.

FIG. 7. *Platanus paucidentata* Dorf. Hypotype no. 6245.

V

VEGETATION AND CLIMATE
DURING HEMPHILLIAN TIME

Chapter V Contents

FIGURES

INTRODUCTION

Comparison of the Mt. Reba, Turlock Lake and Broken Hill floras with others of generally similar age that are now known from central California and adjacent Nevada makes it possible to outline in a tentative manner the distribution of vegetation in the region, the nature of the climates under which they lived, and the elevation of the principal vegetation zones in the central Sierra. Since relatively few floras of Hemphillian age are now known, the analysis is supplemented by a consideration of floras that are slightly older and younger. In the following discussion only the general nature of the vegetation is reviewed, because most of the floras have been treated in monographs elsewhere. Figure 1 shows their geographic occurrence.

CLOSED-CONE PINE FOREST

Closed-cone pine forest dominated the outer coast in the San Francisco Bay region, and ranged more or less continuously southward as judged from the records of the forest elsewhere in central and southern California (Axelrod, 1967a; 1980). Cones of *Pinus lawsoniana*, which are similar to those of the living *P. radiata*, occur in nonmarine beds that lie on Franciscan diabase below the Merced Formation at Mussel Rock, 5 miles (8 km) south of Thornton Beach State Park. Remarkably well preserved cones of *P. lawsoniana* are also in the marine "Merced" Formation on the west side of Pt. Reyes Peninsula. They probably were transported there by a submarine mudflow from nearby land to the east, as inferred from the nature of the sediments. A well-preserved cone of *P. masonii*, which is allied to the living *P. "borealis"* occurs in the lower part of the type Merced Formation south of Thornton Beach State Park (Axelrod, 1967a; 1980). All of these sites are west of the San Andreas fault, and during Hemphillian time were transported northward a few tens of miles (e.g., Crowell, 1962; Dibblee, 1966; Matthews, 1976).

The associated plants in the closed-cone pine forest were chiefly woodland and chaparral taxa, as judged from an analysis of the pollen-poor sedimentary rocks in the lower Merced by Dr. W. S. Ting (unpublished data). Taxa similar to *Quercus wislizenii*, *Q. lobata*, and *Pinus sabiniana* point to a drier and warmer climate in the coastal strip in the Middle Pliocene. This agrees with the records of pollen of *Cercocarpus*, *Crossosoma*, *Dendromecon*, *Forestiera*, *Fraxinus* (cf. *velutina*), and *Prunus* (cf. *ilicifolia*) in the lower Merced Formation. Exotic taxa, including *Carya*, *Juglans*, and *Ostrya*, are present, but their representation is very low. Significantly, there is no indication in the lower part of the Merced section of taxa that represent coast conifer or mixed conifer forest. They gradually assumed dominance in the middle and upper part of the formation as closed-cone pines and their associates were reduced.

LIVE-OAK WOODLAND AND SAVANNA

The Mulholland and Petaluma floras of the San Francisco Bay region are dominated by live-oaks. They also characterize the Etchegoin and Jacalitos floras of the Coalinga region, and dominate the Oakdale and Turlock Lake floras from the lowest piedmont of the central Sierra. Oak woodland and savanna probably extended across the Central Valley of California in the area north of the marine embayment that occupied much of the San Joaquin basin (Fig. 8), inhabiting well-drained sites away from the riparian woodland composed of *Alnus, Persea, Populus, Platanus, Salix,* and *Sapindus.*

In the near-coastal area, woodland included taxa whose closest descendants are now in insular (*Lyonothamnus floribundus, Quercus tomentella*) or coastal (*Ceanothus spinosus, Malosma (Rhus) laurina*) southern California. They indicate mild climate and probably a near-absence of frost, a relation consistent with the wide distribution of avocado (*Persea*) in the region (Fig. 8). The absence from the Oakdale, Etchegoin, and Jacalitos floras of taxa whose nearest relatives are now insular or coastal may have been chiefly the result of warmer summers, and hence higher evaporation with attendant greater water-stress.

The upper limit of oak woodland-savanna on the central Sierran west slope probably was near 1,000-1,500 ft. (305-573 m). This is inferred from the composition of the Oakdale flora, which is dominated by live-oaks and lived at an altitude of not more than 200 ft. (61 m) (Axelrod, 1944). Furthermore, the Mt. Reba flora from an elevation estimated near 2,500 ft. (762 m) was well above the live-oak woodland zone. Moisture there was sufficient to support humid forest species (*Cupressus, Pseudotsuga, Lithocarpus*) that are not recorded in the lowland Oakdale or Turlock Lake floras. The rise in precipitation in later Hemphillian time, as indicated by the Turlock Lake flora, enabled mesic broadleaved sclerophyll vegetation to reach down to lower levels, displacing the more xeric woodland taxa like those recorded in the Oakdale flora. With a further rise in precipitation in the Late Pliocene, the lower margin of conifer forest may have reached the floor of the Central Valley in this area, though evidence of this awaits the discovery of younger floras in this region.

To the northeast, the Verdi flora from the lee of the low Sierran ridge reveals live-oak woodland and savanna at the lower margin of mixed conifer forest (Axelrod, 1958). Conditions in that area were still mild, as judged from the presence there of oaks allied to species (*Quercus chrysolepis, engelmannii, wislizenii*) that are now largely west of the Sierra. Furthermore, the abundant small-leaved cottonwood (*Populus alexanderii*) finds its nearest counterpart only in central California from Santa Cruz southward along the coastal strip, a region of very mild winters. Some summer rain was still in the area, as judged from a Verdi poplar (*P. subwashoensis*) and cottonwood(*P. payettensis*) that are similar to *P. davidiana* of eastern Asia and *P. angustifolia* of the Rocky Mountains, respectively.

OAK-JUNIPER WOODLAND

Taxa that represent this vegetation zone occur in western Nevada in the type Truckee Formation of the Carson Sink area, and in the Esmeralda Formation near Coaldale, 150 miles south. Juniper woodland still inhabits the upper slopes of the ranges in western Nevada, generally above 5,500 to 6,000 ft. (976-1,829 m). However, it is greatly impoverished as compared with the Hemphillian community. The latter vegetation is similar to that now in interior southern California, where walnut-oak woodland meets juniper woodland, as near Pomona and San Bernardino. Relationship is also apparent with the present juniper woodland on the upper desert slopes of the San Gabriel and San Bernardino mountains, where juniper and canyon live-oak (*Quercus chrysolepis*) thrive today. Furthermore, a few rare fossils from the Lower Truckee floras indicate that there probably was forest in the higher hills, a relation recalling interior southern California, where the mixed conifer forest lies above juniper-oak woodland. Precipitation probably was near 20 in. when juniper woodland typified the lowlands of western Nevada. Winters probably had only light frost, to judge from present conditions in southern California where some of the Truckee taxa (*Juglans, Lyonothamnus*) are represented today. Summer rainfall was low, but sufficient for a few relicts (e.g., *Rhus* cf. *lanceolata*) whose nearest descendants are now far to the southeast. A pollen flora from the Wichman beds of early Blancan age (~ 3.5 m.y.), which now occurs at the lower margin of juniper woodland, shows that yellow pine forest displaced woodland vegetation as precipitation increased in the Late Pliocene (Axelrod and Ting, 1960). Similar evidence for the occurrence of forest in the present dry lowlands east of the central and southern Sierra Nevada is also documented by pollen floras from Owens Gorge (3.5 m.y.) and from the somewhat younger (2.7 m.y.) Coso Formation (Axelrod and Ting, 1960). They include a number of taxa now in the *Sequoiadendron* forest on the western flank of the range, implying not only higher rainfall in the lee of the low Sierra, but much milder winters at this time when the range was fully 4,000-6,000 ft. (1,219-1,829 m) lower, depending on latitude. That the pollen in these floras was derived from a forest near at hand, and not from transport across the Sierran barrier, is apparent from the record. There is no evidence in these floras of taxa that contribute to desert, sage, or juniper woodland like that dominating these areas today; they are dominated by forest taxa.

CHAPARRAL

Under this heading is included the occurrence of sclerophyllous shrubs that now commonly form dense brush communities of closely spaced plants. As outlined elsewhere (Axelrod, 1975), chaparral was not a climax community in the Pliocene. Sclerophyllus shrubs were chiefly members of a brush-rich understory of oak woodland vegetation, much as they are today. Although shrubs no doubt dominated local drier sites with thin soil, such communities were seral, and were soon supplanted by woodland in disturbed areas. The widespread brushlands of the present landscape are basically the result of man's diverse activities, notably the clearing of forest and woodland, fire, and overgrazing.

Chaparral taxa attained greatest diversity in central California during the Middle Hemphillian, the driest part of the Tertiary in this region. They are well represented in the Mulholland flora, where there are fossils similar to species now in central California, as well as others that occur today only in southern California, notably *Ceanothus spinosus, Malosma laurina,* and *Schmaltzia* (*Rhus*) *ovata*. In contrast to their good representation in the Mulholland flora, chaparral taxa are scarcely represented in the small Petaluma flora. This may reflect the more distant position of the site of plant accumulation with respect to hill-slopes covered with woodland, where shrubs were presumably well developed. On the other hand, there is a moderate representation of sclerophyllous shrubs in the Oakdale and Turlock Lake floras. At Oakdale, this probably results from the somewhat drier climate as well as the drier substrate there, for the flora lived on a floodplain built up of highly pervious, coarse andesitic detritus. However, the greater prominence of chaparral species in the Turlock Lake flora must be explained in another manner, for it inhabited an area of very high water-table close to a large lake. It may be suggested that the rapid rise in precipitation enabled taxa from the broadleaved sclerophyll belt to invade the lowlands and overlap those of the oak woodland zone, thus increasing the diversity there.

The absence of chaparral taxa in the Mt. Reba flora is consistent with its position on the higher, moister, and cooler slopes of the Sierra, where precipitation was adequate for broadleaved sclerophyll and mixed conifer forests. However, on the drier east side of the range at Verdi, sclerophyllous shrubs (e.g., *Ceanothus*) are present, which is expectable since live-oaks are recorded there at the lower margin of mixed conifer forest. To judge from the occurrence of *Ceanothus* and *Cercocarpus* in the Truckee Formation at sites near Hazen and Sagehen Creek, patches of seral chaparral probably were scattered locally in dry sites provided by volcanic slopes and ridges in the region into the Late Hemphillian ($\sim$ 4 m.y.), after which time forest invaded the region as precipitation increased.

BROADLEAVED SCLEROPHYLL FOREST

This community is well represented in the Mulholland flora of western California. It was composed of several sclerophyllous oaks, madrone, tanbark oak, California laurel, and associated deciduous taxa (*Acer, Alnus, Populus, Prunus, Rosa, Salix*) that inhabited moister sites. It also included taxa whose nearest allies are no longer in the region, notably species of *Acer, Nyssa*, and *Populus* that probaly formed a deciduous understory, as well as species of *Karwinskia, Quercus, Sapindus*, and others that are now in the southwestern United States or Mexico, where there is summer rainfall. Clearly, the Pliocene community was richer than the modern, for it included both evergreen and deciduous taxa that are no longer in the region. The evergreen nature of the community increased as plants that required summer rain gradually were eliminated by the close of the epoch (Axelrod, 1973).

Broadleaved sclerophyll vegetation also inhabited slopes bordering the Petaluma basin not far from the fossil site. Its poorer representation there no doubt results from the smaller sample, not a less diverse community. Collecting is difficult because the plants occur in a vertically-dipping thin shale that is sandwiched between well-cemented thick sandstones. Precipitation in this coastal area probably was not over 25 in. (635 mm), otherwise conifer forest taxa might well be expected in the Mulholland and Petaluma floras, and there is no evidence of them in the record as now known.

Proceeding eastward across the Central Valley to the lower Sierran slope, the Oakdale flora shows that live-oak woodland-savanna, not broadleaved sclerophyll forest, was in the lowest foothill belt. To judge from the nature of the flora, rainfall was near 20 in. (508 mm), and hence too low for broadleaved sclerophyll taxa (e.g., *Quercus* cf. *chrysolepis, Lithocarpus* cf. *densiflorus, Arbutus* cf. *menziesii*). However, sclerophyll forest assumed dominance as rainfall increased upslope, as shown by the Mt. Reba flora. In that area *Lithocarpus* and *Quercus* (cf. *chrysolepis*) are common, and conifers (*Cupressus, Pseudotsuga*) are prominent. They indicate that annual precipitation probably was near 40 in. (1,016 mm) in this area where mixed conifer forest—composed additionally of *Abies*, *Pinus*, and *Sequoiadendron*—replaced sclerophyllous forest as precipitation increased at slightly higher levels to the east. It was near the close of Hemphillian time, when precipitation increased, that broadleaved sclerophyll forest descended to lower altitudes in the range. This is shown by the Turlock Lake flora with its species of *Arbutus*, *Ceanothus*, and *Smilax* which now live at the upper edge of the oak woodland belt, and which in favorable cooler sites also supports local stands of *Pinus ponderosa*.

Broadleaved sclerophyll vegetation was well represented in western Nevada in the Late Miocene, inhabiting warmer slopes bordering *Sequoiadendron* forest, as shown by the Fallon, Chloropagus, Aldrich Station, and Purple Mountain floras (Axelrod, 1956, 1976b). Similar relations are indicated by the Chalk Hills flora (Axelrod, 1962), which has *Arbutus*, *Chrysolepis*, and *Lithocarpus* together with *Abies*, *Pinus*, *Pseudotsuga*, *Sequoiadendron*, and their regular associates that lived in cooler, moister sites. This appears to be the latest record of broadleaved sclerophyll vegetation in that region. Apart from live-oaks, there is no evidence of it in the small Truckee floras to the east. Furthermore, it is not in the Early Hemphillian Mansfield Ranch flora that represents forest (*Abies*, *Pinus*, *Pseudotsuga*, *Mahonia*) in Hungry Valley 25 miles northeast of Reno; only the canyon live-oak *Quercus hannibalii* (cf. *chrysolepis*) represents a surviving relict. Broadleaved sclerophyll forest had been eliminated from the Verdi area by the later Middle Hemphillian, probably because summer evaporation was now too high at the east base of the Sierra Nevada, which was a low risc at this time.

MIXED CONIFER FOREST

This forest does not appear in the record in the coastal strip until Blancan time, as shown by the Napa flora where *Pinus* (cf. *ponderosa*), *Pseudotsuga* (cf. *menziesii*), and their associates are recorded (Axelrod, 1950). The rarity of taxa representing this forest in the Napa flora indicates that it lived in the nearby hills, not in the lowlands where broadleaved sclerophyll vegetation dominated. Somewhat earlier, the lowlands were characterized by live-oak woodland and with broadleaved sclerophyll vegetation in the moister sites, as shown by the Middle Hemphillian Petaluma flora.

The principal contribution that the Mt. Reba flora makes to an understanding of the history of vegetation in the Sierra Nevada is apparent from Figure 18. A mixed conifer forest of *Abies*, *Cupressus*, *Pseudotsuga*, and *Sequoiadendron* blanketed the higher parts of the range, whereas subalpine forest and alpine vegetation dominate the crestal area today. The mixed conifer forest extended southward into the higher southern Sierra, which stood at a higher level in the Pliocene and also in the Miocene (Axelrod, 1957). The presence of mixed forest in the southern Sierra during the Miocene is demonstrated by records of cones of *Pinus prelambertiana* (cf. *lambertiana*) in the Middle Miocene marine sedimentary rocks northeast of Bakersfield. They apparently were transported into the basin from a mountainous area to the east.

A conifer forest of *Pseudotsuga*, *Cupressus*, and associated sclerophylls (*Quercus*, *Lithocarpus*) probably reached down the central west Sierran slope, with its lower margin probably near 1,500-2,000 ft. (457-609 m). Between that level and the Mt. Reba area (alt. ~ 2,500 ft.; 762 m) there probably was a broad transition zone, with evergreen sclerophyll forest occuring chiefly on warmer south-facing slopes. The mixed conifer forest of *Abies*, *Pinus*, and *Sequoiadendron* probably was confined to higher levels

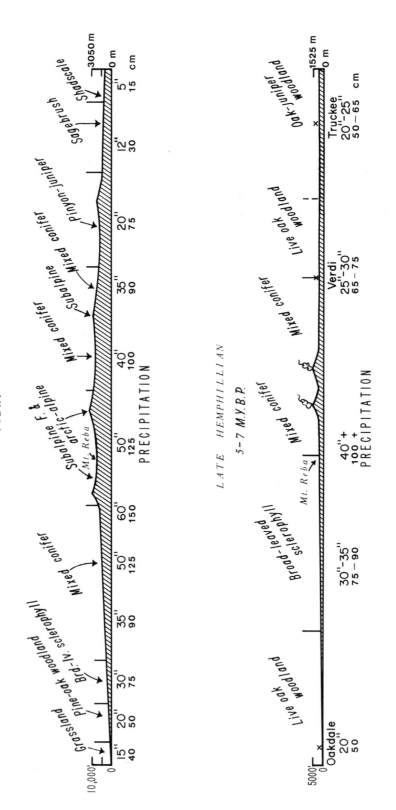

FIG. 18. Comparison of vegetation-topographic profile across the Sierra Nevada and into western Nevada during Hemphillian time and today.

above the Mt. Reba area, to sites where climate was moister and cooler. Mountain juniper (*Juniperus*) evidently was confined to drier, south-facing slopes composed of mudflow breccia and volcanic conglomerate. As outlined elsewhere (Axelrod, 1976a, fig. 4), juniper and other taxa that earlier had a wider distribution in mixed conifer forest zone were confined gradually to higher levels when drier, hotter summers appeared, as a full montane mediterranean climate emerged following the Pliocene.

To the east of the Sierra, the Verdi flora shows that mixed conifer forest was bordered at lower levels by an oak woodland similar to that found today on the lower west slope of the range, where winters are mild. The forest at Verdi does not have conifers of mesic requirements (e.g., *Cupressus*, *Pseudotsuga*, *Sequoiadendron*) like those at Mt. Reba. The Verdi forest of pine (cf. *Pinus attenuata*, *P. ponderosa*, *P. lambertiana*), fir (cf. *Abies concolor*), and their associates indicates drier climate on the east slope. This is consistent with the absence of broadleaved sclerophyll vegetation there, as represented by *Arbutus*, *Chrysolepis* (= *Castanopsis* cf. *chrysophylla*), or *Lithocarpus*, though they were in the area somewhat earlier (Chalk Hills flora) when precipitation was higher. The Verdi forest is rather similar to that in the immediate area today. Pine-fir forest reaches down the canyons on the east front of the Carson Range into the Verdi basin, where precipitation totals 25 in. However, the Verdi flora lived under milder winters, for one of its conifers (*Pinus* cf. *attenuata*), as well as the oaks and a cottonwood, are similar to species that now live west of the Sierra, in central and southern California.

SUBALPINE FOREST?

Since a mixed conifer forest of *Abies*, *Pinus*, *Pseudotsuga*, and *Sequoiadendron* lived chiefly upslope from the Mt. Reba area, the question naturally arises as to whether subalpine forest was in the nearby region farther east. The lower part of the present subalpine zone is a white fir-red fir-Jeffrey pine forest, with pure red fir forest on the cooler, northerly slopes, followed by a mixed forest of red fir-mountain hemlock-white pine, and with whitebark pine appearing at altitudes generally above 8,800-9,000 ft. (2,682-2,743 m) in this sector of the range.

Lower subalpine (e.g., fir) forest may have covered the higher volcanic plugs that rose above the generally even crestline. Among these local areas are Raymond Peak (10,011 ft.; 3,050 m) and Reynolds Peak (9,690 ft.; 2,954 m) near Ebbetts Pass, and Roundtop (10,380 ft.; 3,164 m) near Carson Pass, which now rise about 1,000 ft. (305 m) above the basement of plutonic rocks. As judged from the evidence of uplift suggested by the Mt. Reba flora, these volcanic peaks probably were then about 6,000 ft. (1,829 m) lower, or near 4,000-5,000 ft. (1,219-1,524 m). To the east and southeast of Ebbetts Pass, the volcanic center around Silver Peak (alt. 10,774 ft.; 3,284 m) and Highland Peak (10,934 ft.; 3,369 m) is fully 4,000 ft. (1,219 m) thick. It once surmounted the Sierran crest, and has been downdropped to the east on a major fault that trends down Nobel Canyon. Although this center may have stood somewhat higher than the others, geologic evidence (Curtis, 1951; Wilshire, 1957; Slemmons, 1953, 1966) indicates that all of them were only local eruptive centers, not major stratovolcanos like Mt. Shasta or Mt. Rainier.

The altitude at which temperature would have been appropriate for the lower margin of subalpine forest (*W* 53°F; 11.7°C) and for timberline (*W* 50°F; 10°C) can be estimated from the thermal data deduced for the Mt. Reba flora (*T* 56°F; 13.3°C, *A* 25°F; 13.9°C). As depicted in Figure 19, the thermal level for lower subalpine forest

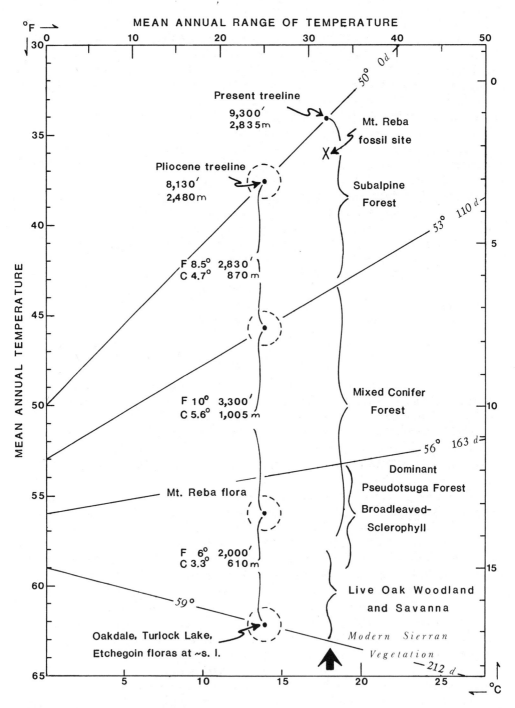

FIG. 19. Estimated elevations needed to support higher vegetation zones in the central Sierra Nevada during Middle Hemphillian time.

represents a mean temperature about 10°F (5.6°C) less than that inferred for the Mt. Reba site. This implies that mountains would either have to be approximately 3,300 ft. (1,006 m) above the Mt. Reba area (10° × 333 ft. = 3,300 ft. or 1,006 m), or have an elevation near 5,300 ft. (1,615 m) above sea level.

This estimate is paralleled by present relations of vegetation and climate in the western Siskiyou Mountains where the lower subalpine forest is at 4,500-5,000 ft. (1,372-1,524 m), depending on exposure. Climate is more equable in this area (*M* 55+) as compared with that in the Sierra (*M* 48) in the ecotone between mixed conifer and subalpine forest. Altitudes of 5,000 ft. (1,524 m) probably were exceeded by the summits of the Silver Peak-Highland Peak volcanic center, but the others appear to have reached this elevation only marginally, to judge from their present thicknesses and small areas. On this basis, most of the higher volcanic peaks would have been fully 2,800 ft. (853 m) below timberline, which is estimated at 8,100 ft. (2,438 m) as compared with 9,300 ft. (2,835 m) today (Fig. 19), when the range of temperature (*A*) is estimated to be 7-8°F (3.9-4.4°C) greater. The higher summits in this part of the range probably did not reach above timberline until the Late Pliocene, as judged from new evidence regarding the age of uplift of the Sierra, as discussed above.

SUMMARY

A comparison of Hemphillian floras in central California and adjacent Nevada provides a provisional basis for outlining the general distribution of vegetation. Closed-cone pine forest was confined to the outer coastal strip, giving way to live-oak woodland in the coastal hills and lowlands to the east. Broadleaved sclerophyll forest dominated at higher, cooler levels in the coastal sector. Oak woodland-savanna covered the present Central Valley area north of the marine embayment, extending up the Sierran piedmont. It was replaced near 1,000-1,500 ft. (305-457 m) by broadleaved sclerophyll vegetation, which probably interfingered with mixed conifer forest near 2,500-3,000 ft. (610-914 m), with the latter extending eastward across the summit divide. Fir forest may have occupied local volcanic centers that rose above the general crestal level, but was not extensive if present. Mixed conifer forest had a drier aspect on the east Sierran slope, with live-oak woodland-savanna occupying drier sites at the lower forest margin in western Nevada. Oak-juniper woodland dominated the lowlands farther east, where relict patches of mixed conifer forest probably were confined to the bordering ranges.

REFERENCES CITED

Axelrod, D. I.
1944   The Oakdale flora. Carnegie Inst. Wash. Pub. 553: 147-166.
1950   A Sonoma florule from Napa, California. Carnegie Inst. Wash. Pub. 590: 23-71.
1956   Mio-Pliocene floras from west-central Nevada. Univ. Calif. Pub. Geol. Sci. 33: 1-316.
1957   Late Tertiary floras and the Sierra Nevadan uplift. Geol. Soc. Amer. Bull. 69: 19-46.
1958   The Pliocene Verdi flora of western Nevada. Univ. Calif. Pub. Geol. Sci. 34: 91-160.
1962   A Pliocene Sequoiadendron forest from western Nevada. Univ. Calif. Pub. Geol. Sci. 39: 195-268.
1967a Evolution of the California closed-cone pine forest. *In* R. N. Philbrick (ed.), Proceedings

of the Symposium on the Biology of the California Islands, pp. 93-150. Santa Barbara, Calif., Botanic Garden.

1967b Geologic history of the Californian insular flora. *In* R. N. Philbrick (ed.), Proceedings of the Symposium on the Biology of the California Islands, pp. 267-315. Santa Barbara, Calif., Botanic Garden.

1973 History of the Mediterranean ecosystem in California. *In* H. A. Mooney and F. di Castri, (eds.), Mediterranean Type Ecosystems: Origin and Structure, pp. 225-277. Springer-Verlag.

1975 Evolution and biogeography of Madrean-Tethyan sclerophyll vegetation. Missouri Bot. Garden Ann. 62: 280-334.

1976a History of the conifer forests, California and Nevada. Univ. Calif. Pub. Botany. 60: 1-62.

1976b Evolution of the Santa Lucia fir (*Abies bracteata*) ecosystem. Missouri Bot. Garden Ann. 63: 24-41.

1980 History of the maritime closed-cone pines, Alta and Baja California. Univ. Calif. Pub. Geol. Sci. 120: 1-143.

Axelrod, D. I., and W. S. Ting
1960 Late Pliocene floras east of the Sierra Nevada. Univ. Calif. Pub. Geol. Sci. 39: 1-118.

Crowell, J. C.
1962 Displacement along the San Andreas fault, California. Geol. Soc. Amer. Spec. Paper 71. 61 pp.

Curtis, G. H.
1951 The geology of the Topaz Lake quadrangle and the eastern half of the Ebbetts Pass quadrangle. Ph.D. thesis, Univ. California, Berkeley. 310 pp.

Dibblee, T. W., Jr.
1966 Evidence for cumulative offset on the San Andreas fault in central and northern California. Calif. Div. Mines and Geol. Bull. 180: 375-384.

Matthews, V., III
1976 Correlation of the Pinnacles and Neeach volcanic formations and their bearing on the San Andreas fault problem. Amer. Assoc. Petrol. Geol. Bull. 60: 2128-2141.

Slemmons, D. B.
1953 Geology of the Sonora Pass region, California. Ph.D. thesis, Univ. California, Berkeley. 201 pp.

1966 Cenozoic volcanism of the central Sierra Nevada, California. Calif. Div. Mines and Geol. Bull. 190: 199-208.

Wilshire, H. G.
1957 The history of Tertiary volcanism near Ebbetts Pass, Alpine County, California. Ph.D. thesis, 126 pp.

VI

SYSTEMATICS

Chapter VI Contents

TAXONOMIC PRINCIPLES

The practice of assigning modern names to fossils was employed by Charles Lyell to date Tertiary marine rocks by the percentage of living mollusks preserved in them. European paleobotanists have followed this lead, applying modern names to the structures of Tertiary plants that seem similar to living taxa. However, the method has not been accepted generally in this country, the chief exception being found in the work of Wolfe (1964, and elsewhere), who has placed a number of previously described Tertiary species in synonymy under modern taxa. Although the fossils described here are scarcely separable from the leaves produced by living plants, they should not be given modern names for several reasons.

It is emphasized that these "species" are known for the most part only from leaf impressions: there is no record of their wood, flowers, the nature of pubescence (its presence and type, or absence), color of the leaves, flowers, or bark; etc. By assigning a distinct name to a Tertiary fossil and noting its modern affinity, we can indicate its relationship without committing ourselves to an opinion of identity that cannot be established. In view of the continuum of taxa in time, it obviously becomes a matter of opinion when a fragment from a tree (a leaf or a winged seed) preserved as a fossil represents a modern species.

Although most living taxa range down into the Pleistocene, there are sufficient differences even between older Pleistocene and modern communities to indicate that they did not live under identical conditions. When we turn to Pliocene and older rocks, there usually is good evidence of close relationship between fossil and living species. However, the Pliocene communities illustrate greater differences between their ecological conditions and those under which the modern exist. Still greater differences in environment are apparent when we compare Miocene or Oligocene floras and related modern forests. Such differences become important when taken in conjunction with evolutionary studies which reveal the dynamic nature of the hereditary process under the influence of changing environment: it militates against the preservation of exact similarity over even brief periods of time. The relations imply that the differences between Tertiary and modern plants, which express different environmental conditions, indicate that the Tertiary species differed physiologically, and hence genetically, from the living ones most like them.

To judge from modern plants, such differences frequently are accompanied by minor morphologic variations (size of plant; nature of bark or root system; color of leaves or cones; color, size, shape, and arrangement of other organs; type of pubescence if present; etc.). These provide useful and valid characters for recognizing living species (or varieties) that occupy different environments, but these features can scarcely be expected to be preserved in a fossil record.

Leaf size is one of the few characters that may reflect important environmental (and presumably genetic) changes. It has provided a basis for designating ecotypic differences between Miocene and Pliocene species. Examples of decreased leaf size in response to diminished rainfall have been described for species of *Acer, Cercocarpus, Fagus, Lyono-thamnus, Populus, Quercus, Rosa, Ulmus,* and other genera. They appear to reflect major environmental differences, to judge from their associates and also from the ecologic occurrence of their living analogues. Thus by assigning separate names to Tertiary species, we imply that they probably were not the same as their nearest descendants, even though the few fossil structures (leaves, seeds, cones) on which these taxa are necessarily based cannot be separated from those of living taxa.

In this regard it is recalled that the leaves of fossil aspen (*Populus pliotremuloides* Axelrod) have been recovered from Late Miocene and Pliocene floras in the lowlands of central California. At Oakdale (Axelrod, 1944c) it is associated with live-oak woodland vegetation, and at Santa Rosa (Axelrod, 1944d) it is found with a rich coast redwood forest and associated laurels (*Persea, Umbellularia*). The living *Populus tremuloides* does not occur anywhere near these environments today, being confined to mountains where below-freezing temperatures with snow and ice regularly occur for at least several weeks every winter. Such conditions are not found in the derivative vegetation with which it is associated in the Neogene. To reconstruct an environment at these fossil sites that would be sufficient to support the living aspen, a terrain of at least 1,000 m would be required, for which there is no supporting geologic evidence. Clearly, the Late Neogene species differed ecologically and genetically from the living *P. tremuloides*, and for this reason was assigned a different name, *P. pliotremuloides*.

As for designating a single leaf, seed, or cone of a Miocene or Oligocene fossil plant by a modern name, this necessarily implies that all the other divergent morphologic characters that distinguish the living species have evolved at the same rate since that time, a conclusion that can scarcely be supported by fossil evidence, and which is clearly demonstrated as false by the precepts of evolution.

Furthermore, applying a modern name to a Miocene fossil is analogous to installing a Chevrolet motor in a Buick, and then calling the automobile a Buick. Granted, it may look like a Buick, but it certainly cannot perform as one. Obviously, if modern California taxa are recognized in the Miocene, they would have little or no paleoecologic or paleoclimatic significance, for they lived under widely different climates and were different physiologically (and genetically).

I am also assigning separate names to these fossil plants because it emphasizes their temporal separation from their nearest living relatives, and thus adds an important time dimension that would be lacking and confusing if living species like limber pine (*Pinus flexilis*), gold-cup oak (*Quercus chrysolepis*), or box elder (*Acer negundo*) were identified in rocks of Tertiary age.

Tertiary paleobotany involves more than taxonomy. It is ecosystematics in the broadest sense of the word, involving the integration of ecology, systematics, and evolution during time (see Duke, 1978). On this basis, the approach I have followed seems reasonable, for it attempts to recognize environmental differences that are expressed by taxa that form parts of chronoclines that reach down into times still largely unknown to us.

TAXONOMIC REVISIONS

During the course of this study, a few changes were made in the disposition of previously described species. These are listed here, with the reasons for the changes. In addition, some changes in the identity of species in other western Neogene floras are noted.

Acer macrophyllum Pursh. Wolfe, 1964, p. N-29, pl. 5, figs 4-6 (Fingerrock flora); and all cited items in synonymy which were earlier placed under *A. oregonianum* Knowlton = *A. oregonianum* Knowlton.

    For reasons, see "Taxonomic Principles," above.

Arbutus trainii MacGinitie. Wolfe, 1964, p. N-31, pl. 5, figs. 8, 9 (Fingerrock flora); pl. 12, figs. 11-14 (Stewart Spring flora) = *A. prexalapensis* Axelrod.

    The leaves figured from the floras in southwestern Nevada, like those from Middlegate flora, are typically smaller than those from the Miocene flora of the Columbia Plateau region, being not long or as broad. They are more nearly related to living Mexican species than to *A. menziesii* Pursh, which is distributed on the Pacific slope from southern California to the southern tip of British Columbia.

Carpinus grandis Unger. Axelrod, 1944e, p. 254, pl. 43, fig. 6 (Alvord Creek flora) = *Juglans alvordensis* Axelrod, n. sp.

    As noted by Wolfe, the Alvord Creek specimen is not a hornbeam. Inasmuch as it is more complete than the leaves from the Stewart Spring flora, it is designated the type of the species.

Chamaecyparis nootkatensis (Lambert) Spach. Wolfe, 1964, p. N-15, pl. 6, figs. 27, 30, 31, 34-37 (Stewart Spring flora) = *C. sierrae* Condit.

    The material illustrated by Wolfe is allied to *C. lawsoniana*, as shown by the stomata on the scale leaves, and by their acute to blunt tips. This species has been collected at a number of localities in western Nevada (Chalk Hills, Purple Mountain, Aldrich Station, Eastgate, Middlegate, Buffalo Canyon, Fingerrock, Golddyke Road). Contrary to Wolfe's statements, these specimens are not similar to *C. nootkatensis*. Thus his notion that the taxon has greatly changed its ecologic requirements since the late Miocene is baseless. Furthermore, its ecologic occurrence at Stewart Spring, and the other Nevada floras of comparable age, parallels that found today in the ecotone between sclerophyll forest and mixed conifer forest, as in the Mount Shasta region. There, *C. lawsoniana* descends as a line of scattered trees along streambanks into oak woodland country, much as it must have done in the Middle and Late Miocene of western Nevada, where it contributed only rarely to the accumulating record. To maintain, as Wolfe has done (1964, 1969), that the Miocene of western Nevada was characterized by a *Quercus-Chamaecyparis* forest is wholly without merit; it rests solely on a lack of understanding of modern ecologic relations.

Holodiscus fryi Wolfe. Wolfe, 1964, p. N-26, pl. 10, figs. 8, 12; fig. 14 (Stewart Spring flora) = *H. idahoensis* Chaney and Axelrod.

    The Stewart Spring material is allied to that of *H. dumosus* (Nuttall) Heller, as is the material from Creede and Thorn Creek. Since they resemble the same modern taxon, as well as one another, they must represent the same species.

Juglans major (Torrey) Heller. Wolfe, 1964, p. N-20, pl. 8, figs. 9, 10 (Stewart Spring flora) = *J. alvordensis* Axelrod, new species (see *Carpinus*, above).

Larix occidentalis Nuttall. Wolfe, 1964, p. N-14, pl. 6, figs. 28, 29 (not fig. 23, which is *L. cedrusensis* Axelrod, n. sp.

    This winged seed of *Larix* is similar to those of the living *L. occidentalis*.

Larix occidentalis Nuttall. Wolfe, 1964, p. N-14, pl. 6, figs. 28, 29 (not fig. 23, which is *L. cedrusensis* Axelrod, see above) (Stewart Spring flora) = *Glyptostrobus oregonensis* Brown.

Lyonothamnus parvifolius (Axelrod) Wolfe. Wolfe, 1964, p. N-26, pl. 10, figs. 1, 14, 15; pl. 11, figs. 1, 3-6; fig. 15 (Stewart Spring flora only) = *L. cedrusensis* Axelrod, n. sp.

    The specimens of *L. parvifolius* occur in the Miocene Aldrich Station, Middlegate, Eastgate, and Buffalo Canyon floras and in the younger Truckee flora near Hazen. All of them differ from the Stewart Spring fossils in having much smaller leaves, and the lobations on the leaflets are about half to two-thirds as broad (basal width 2-3 mm as compared with 4-5 mm). This smaller-leafed species, *C. parvifolius*, may well have been a shrub rather than a tree.

Mahonia.

Howard Schorn (1966) has shown that some species of *Mahonia* have been previously mis-identified, and that others that are in fact distinguishable have been grouped into "large" species. With his approval, some of his revisions of the late Neogene taxa are herein noted and credited to him.

Mahonia malheurensis Arnold. Condit, 1944, p. 46 (Remington Hill flora) = *M. remingtonensis* Schorn.

This species differs from *M. malheurensis* in having more numerous teeth, which are more evenly spaced; a camptodromous venation throughout the length of the blade; and fine, hair-like spines. Among living species, *M. nevinii* (Gray) Fedde shows relationship to the fossil, though the fossil has longer leaflets. The species is allied to *M. subdenticulata* MacGinitie from Florissant.

Mahonia marginata (Lesquereux) Arnold. Axelrod, 1944b, p. 137 (Mulholland flora) (in part) = *M. submarginata* Schorn.

One of the Mulholland leaflets (specimen no. 1644) is oval in outline, has camptodrome sec-ondaries, and is a small leaflet of *submarginata* (see below).

Mahonia marginata (Lesquereux) Arnold. Axelrod, 1944c, p. 163, pl. 33, fig. 6 (Oakdale flora) = *M. submarginata* Schorn.

This species differs from *M. marginata* in its consistently smaller size, and in its pronouncedly attenuated, awl-like teeth. This may reflect adaptation to the drier climate in which it lived, as compared with that in areas of the occurrence of *M. marginata*, which evidently is ancestral to it.

Mahonia marginata (Lesquereux) Arnold. Axelrod, 1944b, p. 137 (Mulholland flora) (in part) = *M. remingtonensis* Schorn.

Two Mulholland species were compared with the leaflets of *M. fremontii*, which are oval as well as lanceolate. It is now apparent that one of the Mulholland leaflets (specimen no. 1643) differs from the lanceolate leaflets of *M. fremontii* in that the teeth are more numerous and the sinuses are shallower.

Mahonia marginata (Lesquereux) Arnold. Axelrod, 1956, p. 295, pl. 29, fig. 2 (Middlegate flora) = *M. middlegateii* Axelrod, n. sp.

This specimen differs from *M. creedensis* Axelrod and from the related *M. marginata* (Les-quereux) Arnold in the much shallower sinuses between the 3-4 marginal teeth, and the tip is not so prominently attenuated.

Mahonia reticulata (MacGinitie) Brown. Axelrod, 1950, p. 60, pl. 3, fig. 1 (Napa flora) = *M.* cf. *re-ticulata* (MacGinitie) Brown.

This appears to be a slender leaflet of this species, but in view of its fragmentary nature it seems best to qualify its affinity.

Odostemon hollicki Dorf. Dorf, 1930, p. 93, pl. 10, fig. 8 only (not fig. 7, which is *Ilex sonomensis* Dorf) (Sonoma flora) = *Mahonia* cf. *reticulata* (MacGinitie) Brown.

This fragmentary specimen is a *Mahonia*, and seems closer to *M. reticulata* than to any other, though it seems best to qualify its identity in view of its fragmentary nature.

Persea coalingensis (Dorf) Axelrod. Axelrod, 1956, p. 297, pl. 28, fig. 5 (Middlegate flora) = *Arbu-tus prexalapensis* Axelrod.

This record of *Persea* is based on a misidentification. The two specimens are *Arbutus*, as in-dicated by the wavering, semi-anastomosing secondaries and the irregular, coarse tertiaries.

Picea breweriana S. Watson. Wolfe, 1964, p. N-14, pl. 6, figs. 4, 5, 8, 9, 13 (Stewart Spring flora) = *Pinus.*

These winged seeds differ from those produced by *P. breweriana* in that the wings are much longer than those in *Picea*, they are not as wide distally, and the entire structure is much larger.

Picea breweriana S. Watson. Wolfe, 1964, p. N-14, pl. 6, figs. 14, 19 only (Stewart Spring flora) = *P. sonomensis* Axelrod.

Pinus florissantii Lesquereux. MacGinitie, 1953, p. 84, pl. 18, fig. 12; pl. 20, figs. 1, 3, 4 (Florissant flora) = *P. sturgisii* Cockerell.

This material is similar to *P. ponderosa*. The cone of *P. florissantii* Lesquereux (MacGinitie, 1953, pl. 19, fig. 2) is allied to *P. flexilis* James, not *ponderosa*.

Pinus florissantii Lesquereux. Axelrod, 1962, p. 227, pl. 42, fig. 9 (Chalk Hills flora) = *P. balfour-oides* Axelrod.

Pinus ponderosa Lawson. Wolfe, 1964, p. N-15, pl. 1, figs. 1, 4 (Fingerrock flora); pl. 8, figs. 32, 33 (Stewart Spring flora) = *Pinus sturgisii* Cockerell.

Pinus wheelerii Cockerell. MacGinitie, 1953, p. 85, pl. 18, fig. 11 only (not figs. 9 and 13, which probably are immature seeds of pine, possibly those of *P. hambachii* Cockerell) (Florissant flora) = *P. florissanti* Lesquereux.

Pinus wheelerii Cockerell. MacGinitie, 1953, p. 85, pl. 18, fig. 10 only (Florissant flora) = *Pseudo-tsuga longifolia* Axelrod.

This winged seed is similar to those from the Trapper Creek flora, and thus represents a genus not previously reported from the Florissant flora. The fossil species seems allied to the Asian *Pseudotsuga forrestii* Craib, a member of the conifer-hardwood forest of the mountains of southern China.

Pinus wheelerii Cockerell. Axelrod, 1962, p. 227, pl. 42, figs. 4-8 (Chalk Hills flora) = *P. balfour-oides* Axelrod.

Pinus leaves MacGinitie, 1953, pl. 18, fig. 3 (Florissant flora) = *Pinus florissantii* Lesquereux.

Populus balsamoides Goeppert. Lesquereux, 1883, p. 248, pl. 55, fig. 4 only (Neroly flora) = *P. gar-berii* Axelrod.

Populus balsamoides Goeppert. Condit, 1938, p. 254, pl. 4, fig. 3; pl. 5, figs. 1, 3 (Neroly flora) = *P. garberii* Axelrod.

Populus prefremontii Dorf. Axelrod, 1944b, pl. 27, fig. 4 only (Mulholland flora) = *P. moragensis* n. sp.

This specimen differs from *P. prefremontii* Dorf in its markedly deltoid shape, acuminate tip, and finely toothed margin, and with teeth distributed along the basal part of the blade. Among living species, it seems allied to *P. dimorpha* Brandegee, a widely distributed tree in northwestern Mexico, notably in Sonora, Sinaloa, and reaching into Chihuahua. Another specimen similar to it was collected by Ralph W. Chaney in 1954; it is herewith added to the Mulholland collection and assigned homeotype no. 6269.

Populus trichocarpa Torrey and Gray. Wolfe, 1964, p. N-18, pl. 8, fig. 3 (Stewart Spring flora: figs. 11 and 12 may be hybrids with *cedrusensis*); and all items in synonymy except *P. alexanderii* Dorf (which remains *P. alexanderii* Dorf) = *P. eotremuloides* Knowlton.

More than one taxon has been included in *P. trichocarpa*. The type of the living species, from near Ventura in southern California, has broadly ovate leaves. This form is chiefly coastal to near-coastal, ranging northward to the San Francisco Bay area, where it is replaced by a form with broadly lanceolate leaves that ranges farther north and occurs also in the mountains of California.

The south-coastal form ("var. *typica*") has leaves similar to those of *P. alexanderii* Dorf, which seems to have been derived from *P. emersonii* Condit from the Miocene Neroly flora. The taxon from cooler northern climates and in the mountains ("var. *montanus*") compares closely with leaves recorded as *P. eotremuloides* Knowlton at many Miocene localities in Nevada, Oregon, and Idaho. It is noteworthy that the small-leaved ovate form, *P. alexanderii* Dorf (cf. "var. *typica*"), from the Pliocene Verdi flora was replaced at that site by the modern *P.*

trichocarpa ("var. *montanus*") as the region was elevated into a colder, more continental climate (see Axelrod, 1958, pl. 19 and pl. 20). Wolfe (1964) maintains that such ecospecies have no significance, though he presents no evidence to support his notion.

Populus washoensis Brown. Axelrod, 1944a, p. 98, pl. 22, figs. 1, 2 (Black Hawk Ranch flora) = *P. garberii* Axelrod.

Prunus sp. Wolfe, 1964, p. N-28, pl. 10, fig. 13 only (Stewart Spring flora) = *Arbutus prexalapensis* Axelrod.

Wolfe identified two specimens as *Prunus*. The one with a serrate margin (pl. 10, fig. 11) is rosaceous and may be *Prunus* or *Rosa*, the photograph not revealing its diagnostic features. However, the specimen illustrated in fig. 13 is certainly not *Prunus*. The margin is entire, and the wavering secondary venation and coarse, irregular mesh are similar to that illustrated by Wolfe's *Arbutus* specimens on pl. 12, figs. 11-14.

The suites of *Arbutus* leaves from the Stewart Spring and Fingerrock floras average much smaller than the material from the Miocene of Oregon and Idaho, and form a derivative taxon, one adapted to a drier, sunnier climate.

Pseudotsuga taxifolia (Lamb.) Britt. Chaney and Mason, 1933, p. 57 (Carpinteria flora) = *P. macrocarpa* (Vasy) Mayr.

Seven wood fragments in the Carpinteria flora were referred to *P. taxifolia* (now = *P. menziesii*) by Chaney and Mason. The basis for this assignment was that "if this species had been living in the immediate vicinity of the deposit, it seems certain that a more complete record would have been left. . . . It seems probable to conclude from the scarcity of its remains . . . that *Pseudotsuga* was not an integral part of the Carpinteria flora, but that the fragments recorded were brought from a distant point, not improbably from what is now Santa Cruz Island . . . where it is known to have been a dominant member of the forest."

That the woody chips could not have drifted to the Carpinteria deposit is evident from the fact that it was not laid down on a beach, but is a stream deposit. Furthermore, there are 6 cones of *P. macrocarpa* in the collections at the Santa Barbara Museum of Natural History and the Los Angeles County Museum. For these reasons, it seems more likely that the wood fragments represent *P. macrocarpa*, not *P. menziesii* (= *taxifolia*). The general rarity of its remains in the deposit probably reflects its occurrence in cooler sites in the nearby hills directly north. From that area a few stragglers probably reached out onto the coastal plain, from which their remains were carried by stream to the area of plant accumulation.

Quercus cedrusensis Wolfe. Wolfe, 1964, p. N-21, pl. 9, fig. 15 (Stewart Spring flora) = *Q. wislizenoides* Axelrod.

The Stewart Spring specimen is no more than a large leaf of this species, which was common in the western Great Basin during the later Miocene. As noted elsewhere, the living form with larger leaves regularly occurs in the broadleaved sclerophyll forest, whereas in the drier oak woodland it produces smaller leaves.

Quercus chrysolepis Liebmann. Wolfe, 1964, p. N-21, pl. 2, figs. 1-10, 14 (Fingerrock flora); pl. 9, figs. 2, 3, 5-7, 12, 16 (Stewart Spring flora) = *Quercus hannibali* Dorf.

Quercus pliopalmerii Axelrod, 1956, pp. 296 and 313 (Mulholland flora) = *Mahonia subsimplex* Schorn.

This oval leaflet, identified initially as *Mahonia* (Axelrod, 1944, p. 137), was erroneously transferred to *Quercus*. The specimen (no. 1644) has camptodrome secondaries and therefore cannot be *Quercus*.

Quercus simulata Knowlton. Condit, 1944, p. 45, pl. 5, fig. 3 (Remington Hill flora) = *Lithocarpus klamathensis* (MacGinitie) Axelrod.

Quercus simulata Knowlton. Axelrod, 1956, p. 291, pl. 13, fig. 12 (Chloropagus flora); pl. 27, figs. 1-4 (Middlegate flora) = *Lithocarpus klamathensis* (MacGinitie) Axelrod.

Quercus simulata Knowlton. Axelrod, 1962, p. 223, pl. 48, fig. 3 (Chalk Hills flora) = *Lithocarpus klamathensis* (MacGinitie) Axelrod.

All of the above-cited records of *Q. simulata* are *Lithocarpus*, differing from *Quercus* in tertiary nervation and in the nature of venation to the marginal teeth. The leaves are similar to those produced by *Lithocarpus* in the inner Coast Ranges and the Sierra Nevada, areas where climate is warmer in summer and colder in winter than in the coastal strip, where the leaves are typically broader.

Rhus alvordensis Axelrod, 1956, p. 305, pl. 30, fig. 8 only (not fig. 9, which remains *R. alvordensis*) (Middlegate flora) = *Sorbus cassiana* Axelrod.

This specimen (no. 4368) is removed from *Rhus* because the secondaries loop well within the margin and supply the marginal teeth with tertiary veins. The leaflets are not biserrate, and hence do not represent *R. alvordensis* Axelrod. This specimen, and others collected subsequently at the Middlegate, Eastgate, and Buffalo Canyon sites, as well as at Stewart Spring and Fingerrock, are more nearly related to Asian taxa, notably *Sorbus acuparia* Linne and *S. pohuashanensis* (Hance) Hedl.

Rhus integrifolia (Nuttall) Bentham and Hooker. Wolfe, 1964, p. N-28, pl. 12, fig. 2 (Stewart Spring flora) = *R. moragensis* Axelrod.

This specimen is similar to the leaves of *R. (Schmaltzia) ovata* in venation. In addition, its leaves are regularly folded like the fossil.

Salix knowltonii Berry. Axelrod, 1956, p. 285, pl. 7, figs. 1, 2 (Aldrich Station); pl. 13, figs. 6, 7 (Chloropagus flora); pl. 19, figs. 1, 2 (Fallon flora); pl. 25, figs. 12, 13 (Middlegate flora); Axelrod, 1962, p. 231; pl. 45, fig. 2; pl. 26, figs. 1-3, 5 (Chalk Hills flora) = *S. storeyana* Axelrod, new name.

Wolfe (1964, p. N-19) pointed out that the type specimen of *S. knowltonii* Berry has sharp teeth, a feature neither described nor figured by Berry (1927). Thus the abundant, entire-margined leaves with numerous looping secondaries that were referred to this species in the Late Miocene Nevada floras require a new name, here chosen as *storeyana*. It comes from Storey County, in which the Chalk Hills flora is located, for at this site the species is represented by abundant, well-preserved material. The leaves of *S. storeyana* Axelrod are similar to those of *S. lemmonii* Bebb, a member of the montane conifer forests of the Sierra Nevada and Klamath Mountains region, ranging northward into Oregon.

Salix truckeana Chaney. Axelrod, 1956, p. 286, pl. 26, fig. 2 (Middlegate flora) = *S. middlegateii* Axelrod, n. sp.

This species has widely spaced, glandular teeth, whereas *S. truckeana* is finely serrate. *S. middlegateii*, as known now from more numerous specimens, resembles the leaves of *S. melanopsis* Nuttall, which ranges from the Sierra Nevada and northern Coast Ranges into British Columbia and to the Rocky Mountains.

Sorbus alvordensis Axelrod. Wolfe, 1964, p. N-28, pl. 5, fig. 3 (Fingerrock flora) = *S. cassiana* Axelrod.

See discussion under *Rhus alvordensis*, above.

Tsuga heterophylla Sargent. Wolfe, 1964, p. N-15, pl. 6, figs. 15, 16, 20, 21, 24 (Stewart Spring flora) = *T. mertensioides* Axelrod.

These seeds obviously are not those produced by a hemlock similar to the modern coastal *T. heterophylla*, but are allied to the montane *T. mertensiana*, which has larger winged seeds.

Undetermined coniferous seed. Wolfe, 1964, pl. 6, fig. 25 (Stewart Spring flora) = *Larix cedrusensis* Axelrod.

This is a small, somewhat aborted winged seed of *Larix*, as indicated by the small wing that tapers distally.

REFERENCES CITED

Axelrod, D. I.
  1944a The Black Hawk Ranch flora. Carnegie Inst. Wash. Pub. 553: 91-100.
  1944b The Mulholland flora. Carnegie Inst. Wash. Pub. 553: 101-146.
  1944c The Oakdale flora. Carnegie Inst. Wash. Pub. 553: 147-166.
  1944d The Sonoma flora. Carnegie Inst. Wash. Pub. 553: 167-206.
  1944e The Alvord Creek flora. Carnegie Inst. Wash. Pub. 553: 225-262.
  1950 A Sonoma flora from Napa, California. Carnegie Inst. Wash. Pub. 590: 23-71.
  1956 Mio-Pliocene floras from west-central Nevada. Univ. Calif. Pub. Geol. Sci. 33: 1-315.
  1958 The Pliocene Verdi flora of western Nevada. Univ. Calif. Pub. Geol. Sci. 34: 91-160.
  1962 A Pliocene *Sequoiadendron* forest from Nevada. Univ. Calif. Pub. Geol. Sci. 39: 195-268.
Berry, E. W.
  1927 The flora of the Esmeralda Formation in Western Nevada. U.S. Nat. Mus. Proc. 72 (art. 23): 1-15.
Chaney, R. W., and H. L. Mason
  1933 A Pleistocene flora from the asphalt deposits at Carpinteria, California. Carnegie Inst. Wash. Pub. 415: 45-79.
Condit, C.
  1938 The San Pablo flora of western central California. Carnegie Inst. Wash. Pub. 476: 217-268.
  1944 The Remington Hill flora. Carnegie Inst. Wash. Pub. 553: 21-55.
Dorf, E.
  1930 Pliocene floras of California. Carnegie Inst. Wash. Pub. 412: 1-112.
Duke, J. A.
  1978 Ecosystematics. Chap. 4, pp. 53-68. *In* J. A. Romberger (ed.), Beltsville Symposia in Agricultural Research 2: Biosystematics in Agriculture. Montclair, N.J.: Allanheld, Osmun.
Lesquereux, L.
  1883 Contributions to the fossil flora of the western territories, Part III. The Cretaceous and Tertiary floras. U.S. Geol. Surv. Terr. Rept. 8. 283 pp.
MacGinitie, H. D.
  1953 Fossil plants of the Florissant beds, Colorado. Carnegie Inst. Wash. Pub. 599: 1-188.
Schorn, H. E.
  1966 Revision of the fossil species of *Mahonia* from North America. M.A. thesis, Univ. California, Berkeley. 150 pp.
Wolfe, J.
  1964 Miocene floras from Fingerrock Wash, southwestern Nevada. U.S. Geol. Surv. Prof. Paper 454-N: N1-N33.
  1969 Neogene floristic and vegetational history of the Pacific Northwest. Madroño 20:83-110.